PRACTICE WORKBOOK

On My Own

Harcourt Brace & Company

Orlando • Atlanta • Austin • Boston • San Francisco • Chicago • Dallas • New York • Toronto • London

http://www.hbschool.com

Copyright © by Harcourt Brace & Company

All rights reserved. No part of this publication may be reproduced or transmitted in any form or by any means, electronic or mechanical, including photocopy, recording, or any information storage and retrieval system, without permission in writing from the publisher.

Permission is hereby granted to individual teachers using the corresponding student's textbook or kit as the major vehicle for regular classroom instruction to photocopy complete pages from this publication in classroom quantities for instructional use and not for resale.

Duplication of this work other than by individual classroom teachers under the conditions specified above requires a license. To order a license to duplicate this work in greater than classroom quantities, contact Customer Service, Harcourt Brace & Company, 6277 Sea Harbor Drive, Orlando, Florida 32887-6777. Telephone: 1-800-225-5425. Fax: 1-800-874-6418 or 407-352-3442.

HARCOURT BRACE and Quill Design is a registered trademark of Harcourt Brace & Company. MATH ADVANTAGE is a trademark of Harcourt Brace & Company.

Printed in the United States of America

ISBN 0-15-307928-2

11 12 13 14   085   05 04 03 02

CONTENTS

Getting Ready for Grade 1

1	One-to-One Correspondence	P4
2	More and Fewer	P5
3	Numbers Through 5	P6
4	Numbers Through 9	P7
5	Ten	P8
6	Greater Than	P9
7	Less Than	P10
8	Order Through 10	P11
9	Ordinal Numbers	P12

CHAPTER 1 — Understanding Addition

1.1	Modeling Addition Story Problems	P13
1.2	Adding 1	P14
1.3	Adding 2	P15
1.4	Using Pictures to Add	P16
1.5	Writing Addition Sentences	P17

CHAPTER 2 — Understanding Subtraction

2.1	Modeling Subtraction Story Problems	P18
2.2	Subtracting 1	P19
2.3	Subtracting 2	P20
2.4	Writing Subtraction Sentences	P21
2.5	Problem Solving • Make a Model	P22

CHAPTER 3 — Addition Combinations

3.1	Order Property	P23
3.2	Addition Combinations	P24
3.3	More Addition Combinations	P25
3.4	Horizontal and Vertical Addition	P26
3.5	Problem Solving • Make a Model	P27

CHAPTER 4 — Addition Facts to 10

4.1	Counting On 1 and 2	P28
4.2	Counting On 3	P29
4.3	Doubles	P30
4.4	Addition Facts Practice	P31
4.5	Problem Solving • Draw a Picture	P32

CHAPTER 5 — Subtraction Combinations

5.1	Subtraction Combinations	P33
5.2	More Subtraction Combinations	P34
5.3	Vertical Subtraction	P35
5.4	Fact Families	P36
5.5	Subtracting to Compare	P37

CHAPTER 6 — Subtraction Facts to 10

6.1	Counting Back 1 and 2	P38
6.2	Counting Back 3	P39
6.3	Subtracting Zero	P40
6.4	Facts Practice	P41
6.5	Problem Solving • Draw a Picture	P42

CHAPTER 7 — Solid Figures

7.1	Solid Figures	P43
7.2	More Solid Figures	P44
7.3	Sorting Solid Figures	P45
7.4	More Sorting Solid Figures	P46
7.5	Problem Solving • Make a Model	P47

CHAPTER 8 — Plane Figures

8.1	Plane Figures	P48
8.2	Sorting Plane Figures	P49
8.3	Congruence	P50
8.4	Symmetry	P51

CHAPTER 9 — Location and Movement

9.1	Open and Closed	P52
9.2	Inside, Outside, On	P53
9.3	Problem Solving • Draw a Picture	P54
9.4	Positions on a Grid	P55

CHAPTER 10 — Patterns

10.1	Identifying Patterns	P56
10.2	Reproducing and Extending Patterns	P57
10.3	Making and Extending Patterns	P58
10.4	Analyzing Patterns	P59

CHAPTER 11 — Addition Facts to 12

- 11.1 Counting On to 12 .... P60
- 11.2 Doubles to 12 .... P61
- 11.3 Three Addends P62
- 11.4 Practice the Facts P63
- 11.5 Problem Solving • Act It Out P64

CHAPTER 12 — Subtraction Facts to 12

- 12.1 Relating Addition and Subtraction P65
- 12.2 Counting Back P66
- 12.3 Compare to Subtract P67
- 12.4 Fact Families P68
- 12.5 Problem Solving • Write a Number Sentence ... P69

CHAPTER 13 — Building Numbers to 100

- 13.1 Tens P70
- 13.2 Tens and Ones to 20 .... P71
- 13.3 Tens and Ones to 50 .... P72
- 13.4 Tens and Ones to 80 .... P73
- 13.5 Tens and Ones to 100 .... P74
- 13.6 Estimating 10 .... P75

CHAPTER 14 — Comparing and Ordering Numbers

- 14.1 Ordinals P76
- 14.2 Greater Than P77
- 14.3 Less Than P78
- 14.4 Before, After, Between P79
- 14.5 Order to 100 .... P80

CHAPTER 15 — Patterns on a Hundred Chart

- 15.1 Counting by Tens P81
- 15.2 Counting by Fives P82
- 15.3 Counting by Two's P83
- 15.4 Even and Odd Numbers P84

CHAPTER 16 — Counting Pennies, Nickels, and Dimes

- 16.1 Pennies and Nickels P85
- 16.2 Pennies and Dimes P86
- 16.3 Counting Collections of Nickels and Pennies ... P87
- 16.4 Counting Collections of Dimes and Pennies ... P88
- 16.5 Problem Solving • Choose the Model .. P89

CHAPTER 17 — Using Pennies, Nickels, and Dimes

- 17.1 Trading Pennies, Nickels, and Dimes P90
- 17.2 Equal Amounts P91
- 17.3 How Much Is Needed? P92
- 17.4 Quarter P93
- 17.5 Problem Solving • Act It Out P94

CHAPTER 18 — Using a Calendar

- 18.1 Ordering Months and Days P95
- 18.2 Reading the Calendar P96
- 18.3 Ordering Events P97
- 18.4 Estimating Time P98

CHAPTER 19 — Telling Time

- 19.1 Reading the Clock P99
- 19.2 Hour P100
- 19.3 Hour P101
- 19.4 Half Hour P102
- 19.5 Problem Solving • Act It Out P103

CHAPTER 20 — Measuring Length

- 20.1 Using Nonstandard Units P104
- 20.2 Measuring in Inch Units P105
- 20.3 Using an Inch Ruler P106
- 20.4 Measuring in Centimeter Units P107
- 20.5 Using a Centimeter Ruler P108

CHAPTER 21 — Measuring Mass, Capacity, and Temperature

- 21.1 Using a Balance P109
- 21.2 Using Nonstandard Units P110
- 21.3 Measuring with Cups P111
- 21.4 Temperature - Hot and Cold P112

CHAPTER 22 — Fractions

- 22.1 Equal and Unequal Parts of Wholes P113
- 22.2 Halves P114
- 22.3 Fourths P115
- 22.4 Thirds P116
- 22.5 Visualizing Results P117
- 22.6 Parts of Groups P118

CHAPTER 23: Organizing Data

- 23.1 Sort and Classify P119
- 23.2 Certain or Impossible P120
- 23.3 Most Likely P121
- 23.4 Tallying Events P122

CHAPTER 24: Making Graphs

- 24.1 Picture Graphs P123
- 24.2 Horizontal Bar Graphs P124
- 24.3 Vertical Bar Graphs P125
- 24.4 Problem Solving • Make a Graph P126

CHAPTER 25: Facts to 18

- 25.1 Doubles Plus One P127
- 25.2 Doubles Minus One P128
- 25.3 Doubles Patterns P129
- 25.4 Doubles Fact Families P130
- 25.5 Problem Solving • Make a Model P131

CHAPTER 26: More About Facts to 18

- 26.1 Make a 10 .... P132
- 26.2 Adding Three Numbers P133
- 26.3 Sums and Differences to 14 .... P134
- 26.4 Sums and Differences to 18 .... P135

CHAPTER 27: Multiply and Divide

- 27.1 Counting Equal Groups P136
- 27.2 How Many in Each Group? P137
- 27.3 How Many Groups? P138
- 27.4 Problem Solving • Draw a Picture P139

CHAPTER 28: Two-Digit Addition and Subtraction

- 28.1 Adding and Subtracting Tens P140
- 28.2 Adding Tens and Ones P141
- 28.3 Subtracting Tens and Ones P142
- 28.4 Reasonable Answer P143

Name _____

GETTING READY 1

One-to-One Correspondence

1. Draw one egg for each chicken.
2. Draw one carrot for each rabbit.
3. Draw one ball for each kitten.
4. Draw one banana for each monkey.

P4 **ON MY OWN**

Name _____

GETTING READY
2

More and Fewer

1.

2.

3.

4.

1.–2. Draw flowers to show more flowers than butterflies.
3.–4. Draw leaves to show fewer leaves than bugs.

ON MY OWN P5

Name _____

Getting Ready 3

Numbers Through 5

1.  0 1 2 3 4 5

2.

   4

3.

   ___

4.

   ___

5.

   ___

6.

   ___

7.

   ___

1. Write the numbers.
2.–7. Count. Write the number that tells how many.

P6 **ON MY OWN**

Name _____

GETTING READY
4

Numbers Through 9

1.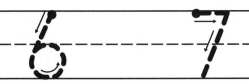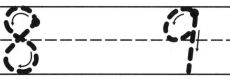

2.

3.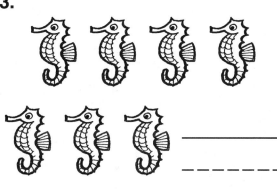

4.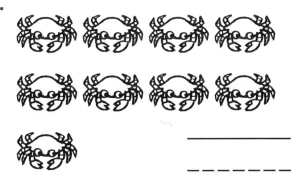

5.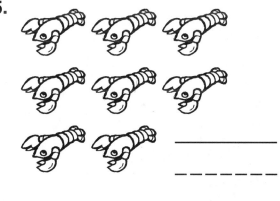

6.

7.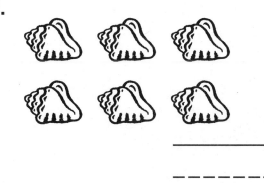

1. Write the numbers.
2.–7. Count. Write the number that tells how many.

ON MY OWN

Name _____

GETTING READY 5

Ten

1.

    10

2.

    _____

3.

    _____

4.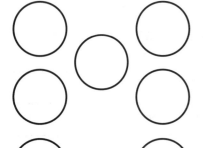

    _____

5.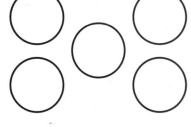

    _____

6.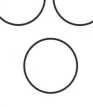

    _____

Write the number that tells how many.

P8 **ON MY OWN**

Greater Than

1.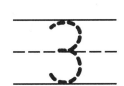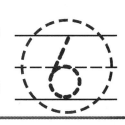

2.    _____   _____

3.   _____   _____

4. 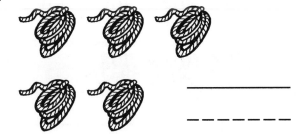 _____ _____

Count. Write the numbers.
Compare the groups. Circle the
number that is greater.

Less Than

1.

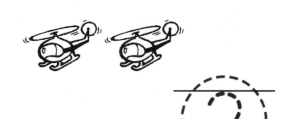

2.

 _____  _____

3.

 _____  _____

4.

 _____  _____

Count. Write the numbers. Compare the groups.
Circle the number that is less.

Order Through 10

GETTING READY 8

1. 4, 5, 6
2. 4, 5
3. 2, 3
4. 1, 3
5. 5, 6
6. 3, 4

Draw circles to show the missing number.
Write the number.

ON MY OWN   P11

Ordinal Numbers

first   second   third   fourth   fifth

1.

2.

3.

4.

5.

1. Color the third animal yellow.
2. Color the fifth animal brown.
3. Color the second animal orange.
4. Color the fourth animal black.
5. Color the first animal red.

Name _____

Modeling Addition Story Problems

LESSON 1.1

Make up a story.
Use cubes to model the story.
Draw the cubes. Write how many there are in all.

1.

____ ____ ____ in all

2.

____ ____ ____ in all

▶ **Problem Solving**

Draw 🐤 and 🐿.
Show one way to make 6.
Write how many there are in all.

3.
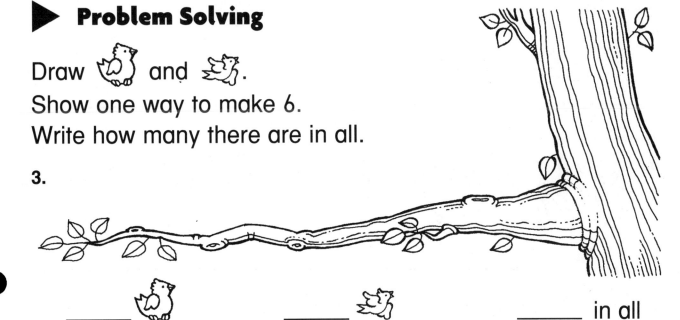
____ ____ ____ in all

ON MY OWN P13

Name _____

LESSON 1.2

Adding 1

▶ **Vocabulary**

Circle the **sum.**

1. 2 + 1 = 3

2. 0 + 1 = 1

Draw and color 1 more. Write the sum.

3.

3 + 1 = 4
       ___
       sum

4.

1 + 1 = ___

5.

5 + 1 = ___

6.

2 + 1 = ___

7.

4 + 1 = ___

8.

3 + 1 = ___

ON MY OWN

Name _____

LESSON 1.3

Adding 2

Draw 2 more balloons.
Color them red.
Write the sum.

1.
3 + 2 = 5

2.
1 + 2 = ____

3.
2 + 2 = ____

4.
4 + 2 = ____

5.
3 + 2 = ____

6.
2 + 2 = ____

▶ **Problem Solving**

Which group has 2 more than 3? Circle the group.

7.

ON MY OWN   P15

Name _____

LESSON 1.4

Using Pictures to Add

Write the sum.

1.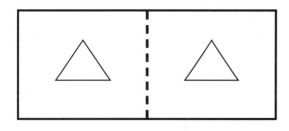

   1 + 1 = __2__

2.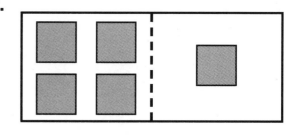

   4 + 1 = ____

3.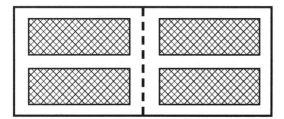

   2 + 2 = ____

4.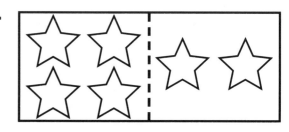

   4 + 2 = ____

5.

   3 + 2 = ____

6.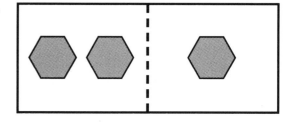

   2 + 1 = ____

7.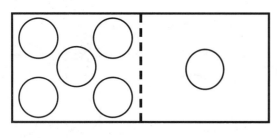

   5 + 1 = ____

8.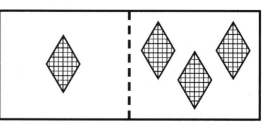

   1 + 3 = ____

Name _____

LESSON 1.5

Writing Addition Sentences

Write the addition sentence.

1.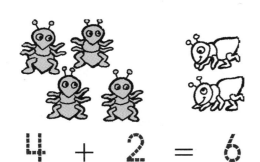

   ___4___ + ___2___ = ___6___

2.

   _____ + _____ = _____

3.

   _____ + _____ = _____

4.

   _____ + _____ = _____

5.

   _____ + _____ = _____

6.

   _____ + _____ = _____

▶ **Problem Solving**

Draw. Write the addition sentence.

7. Draw 1 yellow .
   Draw 2 blue.

   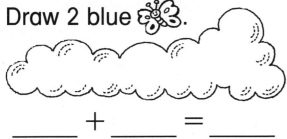

   _____ + _____ = _____

8. Draw 3 orange.
   Draw 1 green.

   _____ + _____ = _____

ON MY OWN   P17

Name _____

Modeling Subtraction Story Problems

LESSON 2.1

Tell a story to a friend.
Use counters to model the story.
Draw the counters. Write how many are left.

1.

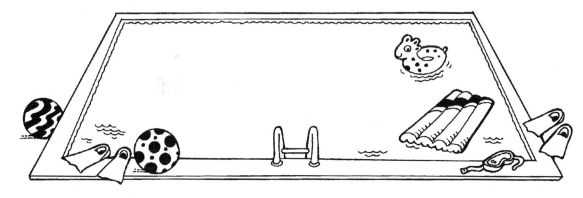

_____ swimmers   _____ go away   _____ are left

2.

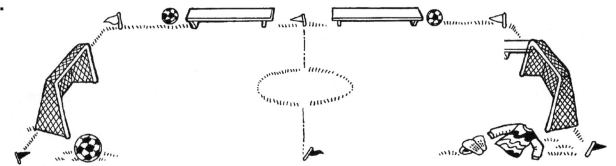

_____ soccer players   _____ go away   _____ are left

3.

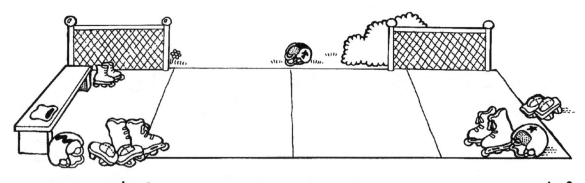

_____ skaters   _____ go away   _____ are left

P18   ON MY OWN

Name _____

LESSON 2.2

Subtracting 1

▶ **Vocabulary**

Circle the **subtraction sentence**.

1. 2 + 1 = 3   2 − 1 = 1

Cross out 1 picture. Write how many are left.

2.   3 − 1 = __2__

3.   4 − 1 = ___

4.   2 − 1 = ___

5.   5 − 1 = ___

▶ **Problem Solving**

Tell a story. Write how many are left.

6.   7.

   5 − 1 = ___   4 − 1 = ___

ON MY OWN P19

Name_____

LESSON 2.3

Subtracting 2

Cross out pictures to show the subtraction sentence. Write how many are left.

1.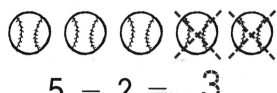

   5 − 2 = __3__

2.

   2 − 1 = ____

3.

   1 − 1 = ____

4.

   5 − 2 = ____

5.

   4 − 2 = ____

6.

   3 − 2 = ____

7.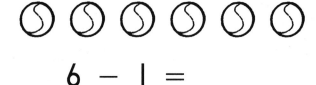

   6 − 1 = ____

8.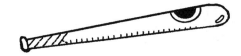

   1 − 1 = ____

9.

   6 − 2 = ____

10.

    4 − 2 = ____

P20 **ON MY OWN**

Name _____

 LESSON 2.4

Writing Subtraction Sentences

▶ **Vocabulary**

Circle the **difference**.

1. 4 − 2 = 2
2. 6 − 2 = 4

Write a subtraction sentence to show the difference.

3.

   _4_ − _2_ = _2_
              difference

4.

   ___ − ___ = ___

5.

   ___ − ___ = ___

6.

   ___ − ___ = ___

7.

   ___ − ___ = ___

8.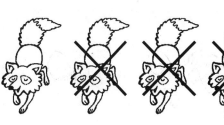

   ___ − ___ = ___

ON MY OWN

Name_____

Problem Solving • Make a Model

LESSON 2.5

Add or subtract. Use counters.
Draw the counters.

1. 4 ducks are swimming.
   I more comes.
   How many in all?

   ___5___ ducks

   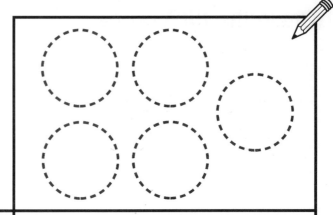

2. 6 kittens are playing.
   4 run away.
   How many are left?

   _____ kittens

3. 3 bees are on a flower.
   2 more come.
   How many in all?

   _____ bees

4. 3 turtles are on a log.
   I goes into the water.
   How many are left?

   _____ turtles

Name_____

LESSON 3.5

Problem Solving • Make a Model

▶ **Vocabulary**

Circle the **cent** sign.

1.   + $ −   ¢

Mark an **X** on the **penny**.

2.

Use pennies to show each price.
Draw them. Write the total amount.

3.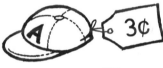

   __7__ ¢

4. 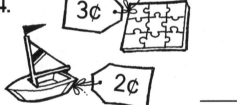   ____ ¢

5. 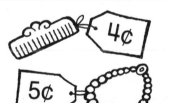   ____ ¢

6.    ____ ¢

Name _____

LESSON 4.1

Counting On 1 and 2

▶ **Vocabulary**

Count on to add.

1. $\begin{array}{r}4\\+1\\\hline 5\end{array}$ *Count on 1.* **5** $\begin{array}{r}7\\+2\\\hline 9\end{array}$ *Count on 2.* **8, 9** $\begin{array}{r}6\\+1\\\hline 7\end{array}$ *Count on 1.* **7**

2. $\begin{array}{r}4\\+2\\\hline\end{array}$ $\begin{array}{r}5\\+1\\\hline\end{array}$ $\begin{array}{r}8\\+2\\\hline\end{array}$ $\begin{array}{r}6\\+1\\\hline\end{array}$ $\begin{array}{r}2\\+1\\\hline\end{array}$

3. $\begin{array}{r}2\\+2\\\hline\end{array}$ $\begin{array}{r}1\\+2\\\hline\end{array}$ $\begin{array}{r}8\\+1\\\hline\end{array}$ $\begin{array}{r}1\\+1\\\hline\end{array}$ $\begin{array}{r}7\\+2\\\hline\end{array}$

4. $\begin{array}{r}5\\+1\\\hline\end{array}$ $\begin{array}{r}7\\+2\\\hline\end{array}$ $\begin{array}{r}4\\+1\\\hline\end{array}$ $\begin{array}{r}2\\+2\\\hline\end{array}$ $\begin{array}{r}6\\+1\\\hline\end{array}$

▶ **Problem Solving**

Tell a story. Write the addition sentence.

5.

6.

P28 **ON MY OWN**

Counting On 3

Count on to add.

1.
$6 + 3 = 9$ (Count on 3. 7, 8, 9)
$5 + 1$
$3 + 2$
$1 + 3$
$5 + 2$

2.
$7 + 3$
$6 + 2$
$3 + 3$
$6 + 1$
$1 + 2$

3.
$4 + 3$
$4 + 2$
$2 + 2$
$2 + 3$
$7 + 1$

4.
$6 + 3$
$5 + 3$
$2 + 1$
$8 + 1$
$7 + 2$

▶ Problem Solving

Solve.

5. Sara has 5 pencils. Tom has 3 pencils. How many pencils do Sara and Tom have in all?

____ pencils

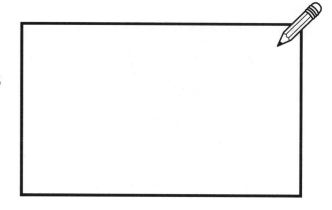

Name _____

LESSON 4.3

Doubles

▶ **Vocabulary**

Circle the **doubles**.

1. $3 + 2 = 5$ $3 + 3 = 6$

Make each picture show a double.
Write the doubles fact.

2.
   $\underline{5} + \underline{5} = \underline{10}$

3.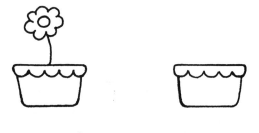
   ___ + ___ = ___

4.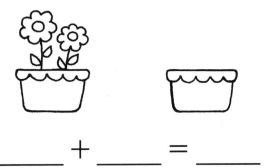
   ___ + ___ = ___

5.
   ___ + ___ = ___

▶ **Problem Solving**

6. Jesse has 3 apples. Matt has double this amount. How many apples does Matt have?

   ___ apples

P30 **ON MY OWN**

Addition Facts Practice

LESSON 4.4

Write the sum. Circle the doubles.

1. $3 + 4 = \underline{7}$ $5 + 1 = \underline{6}$ $(4 + 4 = \underline{8})$

2. $2 + 4 = \underline{}$ $3 + 3 = \underline{}$ $4 + 1 = \underline{}$

3. $0 + 2 = \underline{}$ $6 + 2 = \underline{}$ $1 + 1 = \underline{}$

4. $2 + 2 = \underline{}$ $3 + 2 = \underline{}$ $7 + 1 = \underline{}$

5.
3	1	5	6	6
+1	+1	+5	+2	+3

6.
3	4	2	4	2
+3	+1	+5	+4	+1

▶ **Problem Solving**

7. Doug has 3 cars. Jim has 3 cars. How many cars do Doug and Jim have in all?

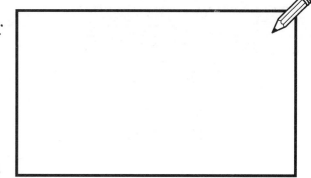

_____ cars

Problem Solving • Draw a Picture

Draw pictures to solve.

1. 4 kittens are white.
   2 kittens are black.
   How many kittens in all?

    6

2. David has 6 yo-yos.
   He gives away 2.
   How many are left?

   _____

3. Jeremy has 3 red balloons.
   Ross has 3 blue balloons.
   How many balloons in all?

   _____

4. Heather has 7 flowers.
   She gives away 2.
   How many are left?

   _____

Name _____

LESSON 5.1

Subtraction Combinations

Use counters. Write ways to subtract.

1. 7 − _7_ = _0_   2. 8 − ___ = ___

3. 7 − ___ = ___   4. 8 − ___ = ___

5. 7 − ___ = ___   6. 8 − ___ = ___

Subtract.

7. 7   6   7   8   7   8
   −4   −2   −7   −4   −6   −3
    3

8. 5   8   7   3   7   7
   −3   −7   −5   −1   −0   −1

9. 6   8   7   8   9   8
   −3   −0   −2   −8   −7   −5

▶ **Problem Solving**

Which answer will be less than 5?
Circle the problem. Solve to check.

10. 8 − 4 = ___   8 − 2 = ___   8 − 1 = ___

Name _____

More Subtraction Combinations

Use counters. Write ways to subtract.

1. 9 − 0 = 9
2. 10 − ___ = ___
3. 9 − ___ = ___
4. 10 − ___ = ___
5. 9 − ___ = ___
6. 10 − ___ = ___
7. 9 − ___ = ___
8. 10 − ___ = ___

9. 9 − ___ = ___
10. 10 − ___ = ___
11. 9 − ___ = ___
12. 10 − ___ = ___
13. 9 − ___ = ___
14. 10 − ___ = ___
15. 9 − ___ = ___
16. 10 − ___ = ___

LESSON 5.2

ON MY OWN

Name _____

LESSON 5.3

Vertical Subtraction

Complete.

1.

   __4__ − __1__ = __3__

2.

   ___ − ___ = ___

3.

   ___ − ___ = ___

▶ Problem Solving

Add or subtract. Circle the answer that is greater.

4. Sam has 7 balloons. ___
   He breaks 2.
   How many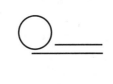
   are left?

   Liz has 3 fish. ___
   She buys 1 more.
   How many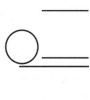
   does she have?

Name _____

LESSON 5.4

Fact Families

▶ **Vocabulary**

Circle the sentence that does not belong in the **fact family**.

1.

$6 + 3 = 9 \quad 9 - 5 = 4$
$3 + 6 = 9 \quad 9 - 6 = 3$
$9 - 3 = 6$

Add or subtract. Write the numbers in the fact family.

2.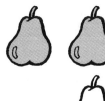

$\begin{array}{r}4\\+2\\\hline 6\end{array}\quad\begin{array}{r}2\\+4\\\hline 6\end{array}\quad\begin{array}{r}6\\-2\\\hline 4\end{array}\quad\begin{array}{r}6\\-4\\\hline 2\end{array}$

[4] [2] [6]

3.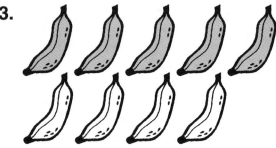

$\begin{array}{r}5\\+4\\\hline\end{array}\quad\begin{array}{r}4\\+5\\\hline\end{array}\quad\begin{array}{r}9\\-4\\\hline\end{array}\quad\begin{array}{r}9\\-5\\\hline\end{array}$

☐ ☐ ☐

4.

$\begin{array}{r}5\\+3\\\hline\end{array}\quad\begin{array}{r}3\\+5\\\hline\end{array}\quad\begin{array}{r}8\\-3\\\hline\end{array}\quad\begin{array}{r}8\\-5\\\hline\end{array}$

☐ ☐ ☐

▶ **Problem Solving**

5. Tell a story. Write the numbers in the fact family.

   ☐ ☐ ☐

Name _____

LESSON 5.5

Subtracting to Compare

Draw lines to match. Subtract to compare.
Write how many more.

1.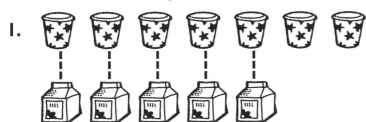

   7 − 5 = __2__

   __2__ more

2.

   6 − 2 = ____

   ____ more

3.

   5 − 4 = ____

   ____ more

4.

   4 − 1 = ____

   ____ more

▶ **Problem Solving**

Solve.

5.  You have 8 🌼. ____

   You have 6 🌷. −____

   How many more
   🌼 do you have? ____

6.  You have 6 🔒. ____

   You have 1 🔑. −____

   How many more
   🔑 do you need? ____

Name_____

LESSON 6.1

Counting Back 1 and 2

▶ **Vocabulary**

Use the number line.
Count back to subtract.

1.

    $8 - 2 = \underline{6}$

    Start at 8. Count back 2. Where are you?

2.
    $5 - 1 = \underline{}$

3.
    $3 - 2 = \underline{}$

4.
    $4 - 2 = \underline{}$

5.
    $4 - 1 = \underline{}$

6.
    $10 - 2 = \underline{}$

7.
    $9 - 2 = \underline{}$

▶ **Problem Solving**

Solve.

8. Sophie had 5 balloons. 3 blew away. How many balloons does she have left?

    _____ balloons

P38 **ON MY OWN**

Name _____

LESSON 6.2

Counting Back 3

Use the number line. Count back to subtract.

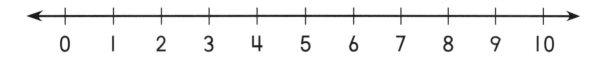

1. $8 - 3 = \underline{5}$ $9 - 3 = \underline{}$ $6 - 3 = \underline{}$

2. $4 - 2 = \underline{}$ $7 - 1 = \underline{}$ $9 - 2 = \underline{}$

3. $3 - 1 = \underline{}$ $3 - 3 = \underline{}$ $2 - 1 = \underline{}$

4.
$$\begin{array}{c}7\\-3\\\hline\end{array}\quad\begin{array}{c}8\\-1\\\hline\end{array}\quad\begin{array}{c}4\\-2\\\hline\end{array}\quad\begin{array}{c}8\\-3\\\hline\end{array}\quad\begin{array}{c}7\\-1\\\hline\end{array}\quad\begin{array}{c}4\\-3\\\hline\end{array}$$

5.
$$\begin{array}{c}5\\-2\\\hline\end{array}\quad\begin{array}{c}2\\-1\\\hline\end{array}\quad\begin{array}{c}3\\-3\\\hline\end{array}\quad\begin{array}{c}6\\-2\\\hline\end{array}\quad\begin{array}{c}4\\-3\\\hline\end{array}\quad\begin{array}{c}6\\-1\\\hline\end{array}$$

▶ **Problem Solving**

Solve.

6. Use these numbers. Write a subtraction sentence.

____ − ____ = ____

ON MY OWN P39

Name _____

LESSON 6.3

Subtracting Zero

Subtract. Circle all the zero facts.

1.  $\begin{array}{r} 6 \\ -6 \\ \hline 0 \end{array}$ $\begin{array}{r} 6 \\ -0 \\ \hline 6 \end{array}$

2. $\begin{array}{r} 9 \\ -9 \\ \hline \end{array}$ $\begin{array}{r} 5 \\ -0 \\ \hline \end{array}$ $\begin{array}{r} 8 \\ -3 \\ \hline \end{array}$ $\begin{array}{r} 5 \\ -5 \\ \hline \end{array}$ $\begin{array}{r} 3 \\ -3 \\ \hline \end{array}$ $\begin{array}{r} 7 \\ -3 \\ \hline \end{array}$

3. $\begin{array}{r} 7 \\ -2 \\ \hline \end{array}$ $\begin{array}{r} 8 \\ -8 \\ \hline \end{array}$ $\begin{array}{r} 5 \\ -4 \\ \hline \end{array}$ $\begin{array}{r} 4 \\ -4 \\ \hline \end{array}$ $\begin{array}{r} 5 \\ -1 \\ \hline \end{array}$ $\begin{array}{r} 3 \\ -0 \\ \hline \end{array}$

4. $\begin{array}{r} 4 \\ -1 \\ \hline \end{array}$ $\begin{array}{r} 7 \\ -7 \\ \hline \end{array}$ $\begin{array}{r} 3 \\ -2 \\ \hline \end{array}$ $\begin{array}{r} 10 \\ -1 \\ \hline \end{array}$ $\begin{array}{r} 2 \\ -1 \\ \hline \end{array}$ $\begin{array}{r} 9 \\ -9 \\ \hline \end{array}$

▶ **Problem Solving**

Write the subtraction sentence.

5.

___ − ___ = ___

Name _____

LESSON 6.4

Facts Practice

Add or subtract. Fill in the numbers.

1.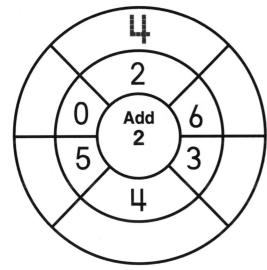

2.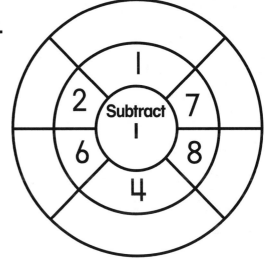

3.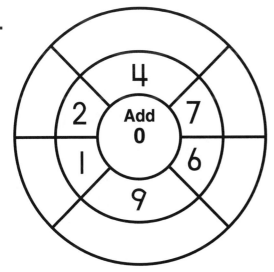

4.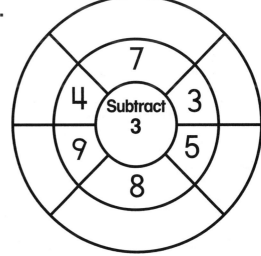

▶ Problem Solving

Circle **add** or **subtract**. Solve.

5. Carol has 6 flowers.
   She picks 4 more.
   How many flowers does she have in all? _____ flowers

        add   **subtract**

ON MY OWN P41

Problem Solving • Draw a Picture

Add or subtract.
Draw more things, or cross things out.

1. 3 buckets are in a row.
   Make 2 more.
   How many now? __5__

2. 6 starfish are on the beach.
   2 swim away.
   How many now? _____

   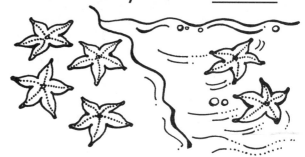

3. 1 chair is in the sand.
   Make 1 more.
   How many now? _____

4. 4 seagulls are walking.
   1 flies away.
   How many now? _____

5. 5 puppies are sleeping.
   3 run away.
   How many now? _____

6. 2 balls are in the yard.
   Make 2 more.
   How many now? _____

Name _____

Problem Solving • Make a Model

Build the model. Write how many cubes you used.

1.

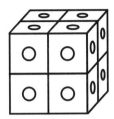

____8____ cubes

2.

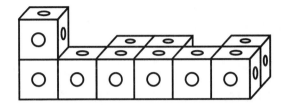

_____ cubes

3.

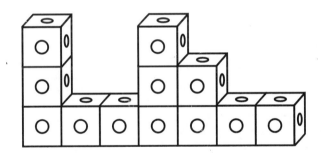

_____ cubes

4.

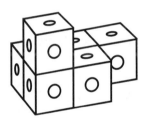

_____ cubes

5.

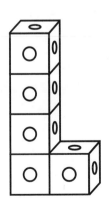

_____ cubes

6.

_____ cubes

Name_____

LESSON 8.1

Plane Figures

▶ **Vocabulary**

Color each **flat face**.

1.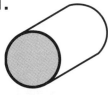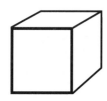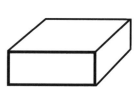

    circle   square   triangle   rectangle

2. Color the triangles.

3. Color the squares.

4. Color the circles.

               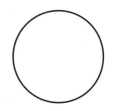

▶ **Problem Solving**

5. Color the figure that has all **flat faces**.

   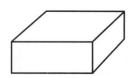   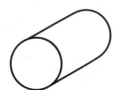

ON MY OWN

Name _____

LESSON 8.2

Sorting Plane Figures

Trace each side. [blue]

Draw a ◯ on each corner. [red]

Write how many sides and corners.

1.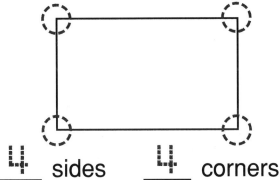

   __4__ sides   __4__ corners

2.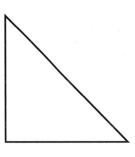

   ___ sides   ___ corners

3.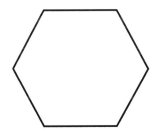

   ___ sides   ___ corners

4.

   ___ sides   ___ corners

▶ **Problem Solving**

5. Circle the figure that has 4 corners and 4 sides.

         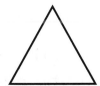

6. Circle the figure that has 5 corners and 5 sides.

      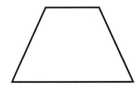

ON MY OWN

Name _____

LESSON 8.3

Congruence

Color the figures that are the same size and shape.

1.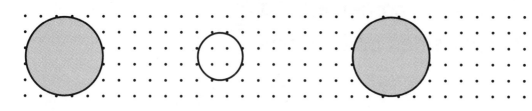

2.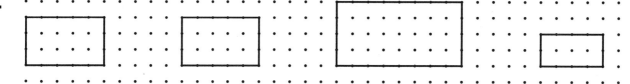

3.

4.

 Problem Solving

Circle the figure.

5. I have no corners.
   I have no sides.
   What am I?

6. I have 4 corners.
   I have 4 sides.
   What am I?

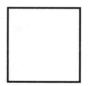

Name _____

LESSON 8.4

Symmetry

Draw a line to make two sides that match.

1.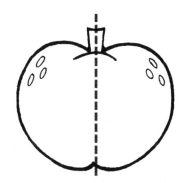

2.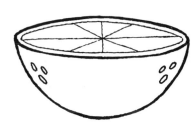

3.

4.

5.

6.

▶ **Problem Solving**

7. Circle the figure with two parts that do not match.

ON MY OWN P51

Name_____

  LESSON 9.1

Open and Closed

▶ **Vocabulary**

1. Circle the **open** figure.

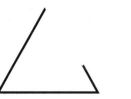 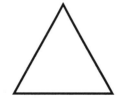

2. Circle the **closed** figure.

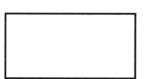 

Color each closed figure. Circle each open figure.

3.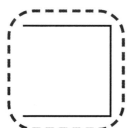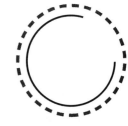

4.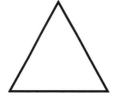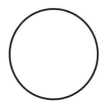

5.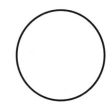

▶ **Problem Solving**

Circle the letters that are open figures.

6. E B R S D C

Name _____

LESSON 9.2

Inside, Outside, On

▶ **Vocabulary**

Color the squares.

1.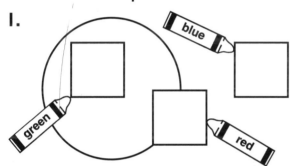

2.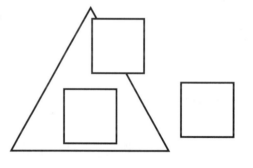

3.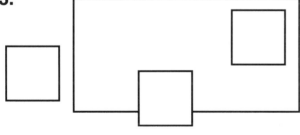

4.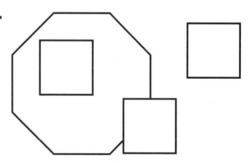

▶ **Problem Solving**

5. Which shape is inside both circles? Color the shape.

ON MY OWN   P53

Problem Solving • Draw a Picture

LESSON 9.3

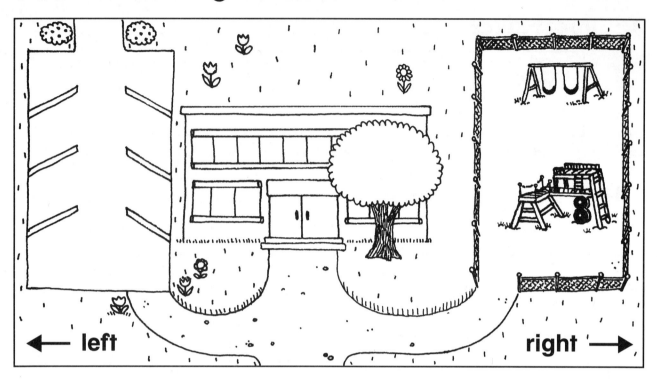

← left   right →

Draw to complete the map.

1. Draw a 🚗 to the left of the 🏫.

2. Draw a 📪 to the right of the 🌳.

3. Draw two 👦 in the ▦.

Circle **left** or **right**.

4. You walk to the 🚗 from the 🏫.

   Which way are you going?

   left   right

5. You go from the 🌳 to the 📪.

   Which way are you going?

   left   right

P54 **ON MY OWN**

Name _____

LESSON 9.4

Positions on a Grid

Start at ☆. Follow directions to draw shapes on the grid.

1.

Right	Up	Draw
5	6	○
2	7	▭

Right	Up	Draw
4	3	△
1	4	▭

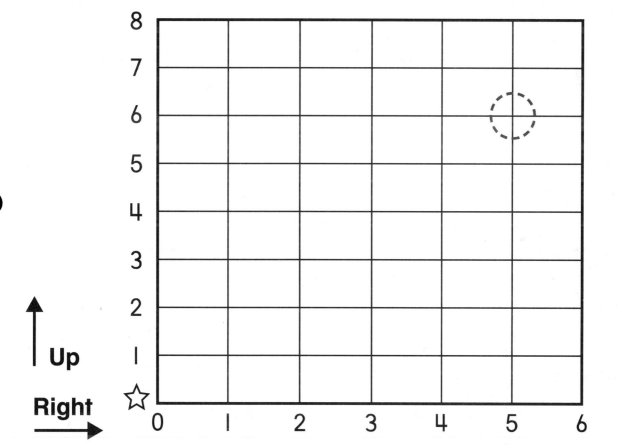

▶ Problem Solving

Look at the grid. Circle the correct shape.

2. What shape is to the left of the triangle?

3. Which shape is farther to the right?

Name_____

LESSON 10.1

Identifying Patterns

Color the R stars ▪red▪.
Color the B stars ▪blue▪.
Color the Y stars ▪yellow▪.
Read the pattern. Then color to continue it.

1.

2.

3.

4.

▶ **Problem Solving**

5. Tina drew these shapes.

   Alan drew these shapes.

   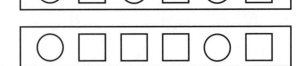

   Circle the shapes that show a pattern.

P56 **ON MY OWN**

Name _____

Reproducing and Extending Patterns

LESSON 10.2

Color the squares to copy and continue the pattern.

1.

red	blue	red	blue	red	blue

red	blue	red	blue	red	blue		

2.

blue	red	blue	red	blue	red

3.

red	red	blue	red	red	blue

▶ **Problem Solving**

4. Make your own pattern. Use red and blue crayons.

5. Write a number to continue the pattern.

3	4	5	3	4	5	___

ON MY OWN P57

Making and Extending Patterns

LESSON 10.3

Use shapes to continue the pattern.
Then use the same shapes to make a different pattern.
Draw the shapes to show your new pattern.

1.

2.

3.

▶ **Problem Solving**

Draw the missing shapes to complete the pattern.

4. ___ ___

5.  ___  ___

Name _____

LESSON 10.4

Analyzing Patterns

Find the mistake in the pattern. Cross it out.
Then use shapes to show the pattern the correct way.
Draw and color the shapes.

1.

2.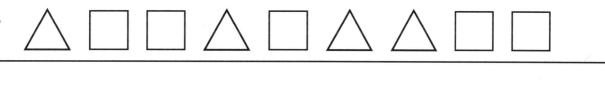

3.

4.

▶ **Problem Solving**

5. Write the numbers to continue the pattern.
   0 1 2 0 1 2 0 1 2 ___ ___ ___

6. Use 6, 7, and 8 to make your own number pattern.

   ___ ___ ___ ___ ___ ___

Name _____

LESSON 11.1

Counting On to 12

Circle the greater number. Count on to add.

1. 1   9   6   3   7
   +(8)   +2   +3   +2   +1
   ───   ───   ───   ───   ───
    9

2. 9   2   8   5   9
   +1   +4   +2   +3   +3
   ───   ───   ───   ───   ───

3. 8   3   2   3   1
   +3   +9   +5   +2   +9
   ───   ───   ───   ───   ───

4. 2 + 7 = ___   7 + 3 = ___   8 + 2 = ___

5. 5 + 2 = ___   3 + 6 = ___   2 + 3 = ___

6. 3 + 9 = ___   4 + 3 = ___   1 + 7 = ___

▶ **Problem Solving**

7. Martha had 5 dolls. She got 2 more. How many does she have in all? ___ dolls

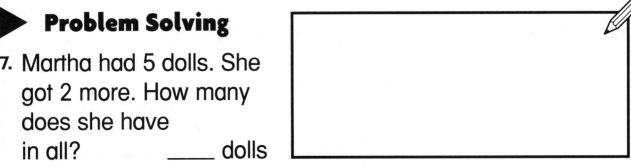

P60 **ON MY OWN**

Name _____

LESSON 11.2

Doubles to 12

Write the sums. Circle the doubles.

1. (4 + 4) = 8   6 + 4 = ___   5 + 5 = ___

2. 6 + 2 = ___   6 + 6 = ___   7 + 1 = ___

3. 7 + 2 = ___   3 + 3 = ___   5 + 2 = ___

4.
```
  5      5      4      2      9
+ 3    + 5    + 2    + 2    + 2
```

5.
```
  3      8      6      3      4
+ 3    + 2    + 6    + 6    + 4
```

▶ **Problem Solving**

6. Caleb spent 6¢. John spent 6¢. How much did they spend in all? ____ ¢

ON MY OWN P61

Name _____

LESSON 11.3

Three Addends

Use cubes. Write the sum.

1.
$$\begin{array}{r}4\\5\\+0\\\hline 9\end{array}\quad \begin{array}{r}4\\3\\+1\\\hline\end{array}\quad \begin{array}{r}6\\6\\+0\\\hline\end{array}\quad \begin{array}{r}4\\5\\+2\\\hline\end{array}\quad \begin{array}{r}6\\1\\+4\\\hline\end{array}$$

2.
$$\begin{array}{r}5\\3\\+2\\\hline\end{array}\quad \begin{array}{r}3\\2\\+3\\\hline\end{array}\quad \begin{array}{r}3\\4\\+3\\\hline\end{array}\quad \begin{array}{r}2\\5\\+5\\\hline\end{array}\quad \begin{array}{r}2\\3\\+4\\\hline\end{array}$$

3.
$$\begin{array}{r}2\\1\\+6\\\hline\end{array}\quad \begin{array}{r}3\\3\\+4\\\hline\end{array}\quad \begin{array}{r}3\\5\\+2\\\hline\end{array}\quad \begin{array}{r}1\\1\\+6\\\hline\end{array}\quad \begin{array}{r}2\\5\\+1\\\hline\end{array}$$

4.
$$\begin{array}{r}7\\3\\+2\\\hline\end{array}\quad \begin{array}{r}4\\2\\+2\\\hline\end{array}\quad \begin{array}{r}1\\3\\+2\\\hline\end{array}\quad \begin{array}{r}7\\4\\+1\\\hline\end{array}\quad \begin{array}{r}5\\2\\+1\\\hline\end{array}$$

▶ **Problem Solving**

Circle the addition sentence that you think has the greater sum. Solve to check.

5. $2 + 4 + 1 =$ _____
   $5 + 1 + 2 =$ _____

6. $4 + 4 + 3 =$ _____
   $5 + 0 + 5 =$ _____

P62 **ON MY OWN**

Name _____

LESSON 11.4

Practice the Facts

Add. Color green the trees that have a sum of 10, 11, or 12.

1.

 4 7 +1 = 12 ; 7 +2 ; 2 5 +2 ; 3 +5 ; 6 +6

2.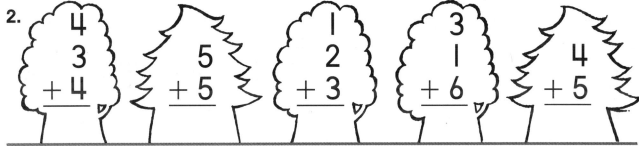

 4 3 +4 ; 5 +5 ; 1 2 +3 ; 3 1 +6 ; 4 +5

3.

 2 +5 ; 8 +4 ; 5 +6 ; 2 7 +3 ; 6 +3

4.

 7 +3 ; 8 +1 ; 6 +5 ; 3 4 +3 ; 7 2 +2

▶ **Problem Solving**

Circle the addition sentences that are correct.

5. 6 + 6 = 12   5 + 4 = 10

   3 + 8 = 12   7 + 2 = 9

ON MY OWN   P63

Name_____

LESSON 11.5

Problem Solving • Act It Out

Act it out. Write the number sentence.

1.
5 ducks swam in the pond.
4 more ducks came to swim.
How many were swimming?

__5__ + __4__ = __9__

__9__ ducks

2.
6 cats played on the rug.
6 cats played on the bed.
How many were playing?

____ + ____ = ____

____ cats

3.
3 goats ran on the hill.
4 goats ran in the field.
How many were running?

____ + ____ = ____

____ goats

4.
7 white rabbits were eating.
3 brown rabbits were eating.
How many were eating?

____ + ____ = ____

____ rabbits

5.
3 frogs slept on a log.
8 frogs slept on a rock.
How many were sleeping?

____ + ____ = ____

____ frogs

6.
4 squirrels sat on a fence.
4 more squirrels came to sit.
How many were sitting?

____ + ____ = ____

____ squirrels

7. Which number sentence goes with the story? Circle it.

3 turtles were walking.   3 + 4 + 2 = 9
4 turtles were sleeping.
2 turtles were swimming.   3 + 4 + 1 = 8
How many turtles were there in all?   5 + 2 + 2 = 9

P64 **ON MY OWN**

Name _____

LESSON 12.1

Relating Addition and Subtraction

Add. Then subtract.

1.

   $8 + 4 = \underline{12}$

   $12 - 4 = \underline{8}$

2.

   $7 + 6 = \underline{}$

   $13 - 6 = \underline{}$

3.
6	10	7	9	9	10
+4	−4	+2	−2	+1	−1

4.
5	8	9	11	8	10
+3	−3	+2	−2	+2	−2

▶ **Problem Solving**

5. Two numbers are added. The sum is 8. One number is 5. What is the other number? _____

6. Two numbers are added. The sum is 6. One number is 4. What is the other number? _____

ON MY OWN   P 65

Name _____

LESSON 12.2

Counting Back

▶ **Vocabulary**

Count back to subtract.
Use the number line if you need it.

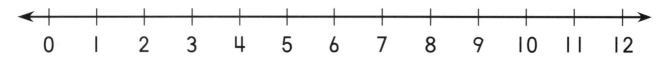

1. 11   9   12   7   6
   −3 −1 −2 −2 −1
   ───
    8

2. 10   7   9   8   12
   −2 −1 −3 −1 −3

3.   8 11   5   4   10
   −2 −2 −1 −3 −3

▶ **Problem Solving**

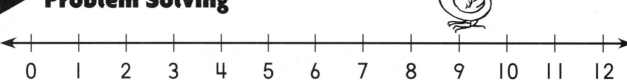

4. A number line was painted on the playground.
   A bird was standing on number 9.
   It took 3 hops back. What number was it on? _____

P66 **ON MY OWN**

Name _____

LESSON 12.3

Compare to Subtract

Compare. Then subtract.

1.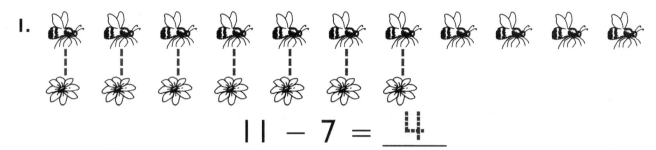

   $11 - 7 = \underline{4}$

2.

   $10 - 8 = \underline{}$

3.

   $11 - 9 = \underline{}$

4.

   $12 - 8 = \underline{}$

▶ **Problem Solving**

5. Ashley has 12 flowers. Cody has 9 flowers. How many fewer flowers does Cody have?

   _____ fewer flowers

ON MY OWN P67

Name_____

Fact Families

LESSON 12.4

Add and subtract.

1. 6 + 4 = 10
    4 + 6 = 10
    10 − 4 = 6
    10 − 6 = 4

2. 3 + 4 = ___
    4 + 3 = ___
    7 − 4 = ___
    7 − 3 = ___

3.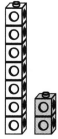
    7 + 2 = ___
    2 + 7 = ___
    9 − 2 = ___
    9 − 7 = ___

4. 5 + 6 = ___
    6 + 5 = ___
    11 − 6 = ___
    11 − 5 = ___

5.
    8 + 1 = ___
    1 + 8 = ___
    9 − 1 = ___
    9 − 8 = ___

6. 4 + 2 = ___
    2 + 4 = ___
    6 − 2 = ___
    6 − 4 = ___

▶ **Problem Solving**

7. Write or draw a story problem that uses these numbers.

   7   3   10

   Write the number sentence your story shows.

   ___ ◯ ___ = ___

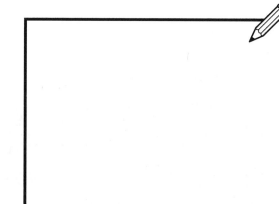

Name _____

LESSON 12.5

Problem Solving • Write a Number Sentence

Write the number sentence the story problem shows.

1. There are 10 dogs.
   There are 7 cats.
   How many more dogs than cats are there?

    __3__ more dogs

   10 ⊖ 7 = 3

2. There are 9 goldfish.
   There are 7 guppies.
   How many more goldfish than guppies are there?

    ____ more goldfish

   ___ ◯ ___ = ___

3. Jenny had 6 cookies.
   She ate 3 of them.
   How many cookies does she have left?

    ____ cookies

   ___ ◯ ___ = ___

4. Jason had 2 apples.
   His grandmother gave him 3 more apples.
   How many apples does he have in all?

    ____ apples

   ___ ◯ ___ = ___

ON MY OWN P69

Name _____

LESSON 13.1

Tens

Write how many tens. Write the number.

1.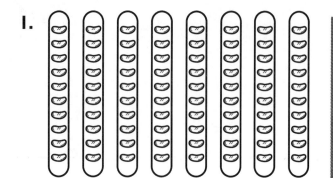

   __8__ tens = __80__
   eighty

2.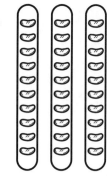

   ____ tens = ____
   thirty

3.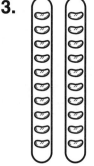

   ____ tens = ____
   twenty

4.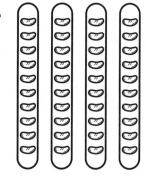

   ____ tens = ____
   forty

▶ **Problem Solving**

5. Circle the box that has 60.

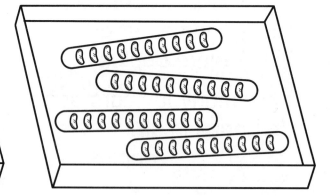

Name _____

LESSON 13.2

Tens and Ones to 20

▶ **Vocabulary**

1. Circle the **tens.**

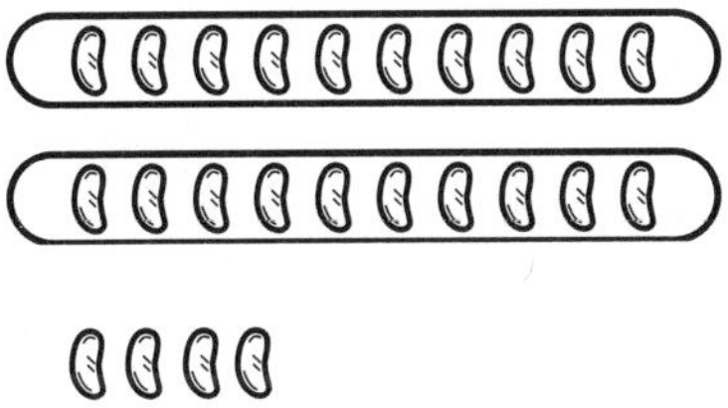

2. Circle the **ones.**

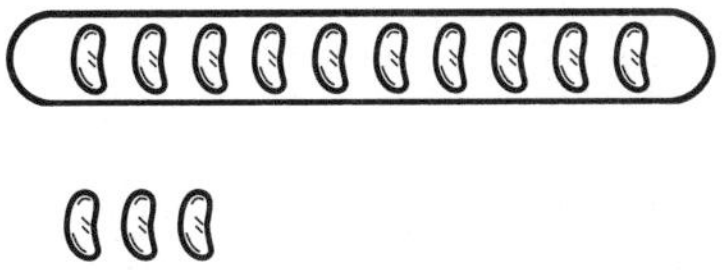

Write how many tens and ones. Write the number.

3. 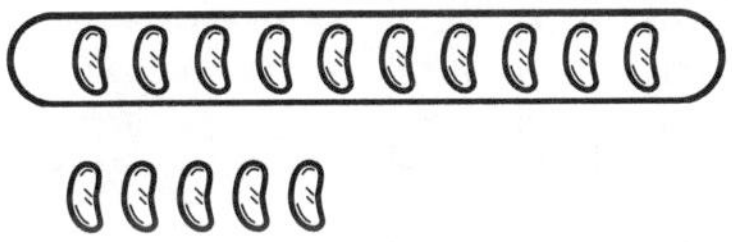 __1__ ten __5__ ones = __15__

4. 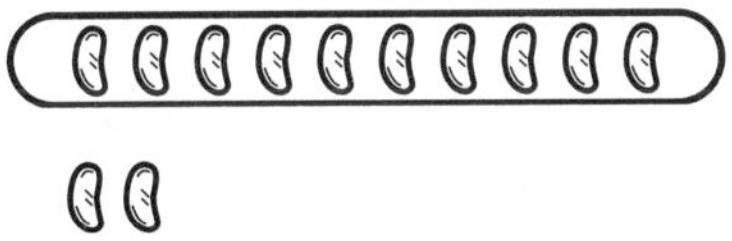 ____ ten ____ ones = ____

5. 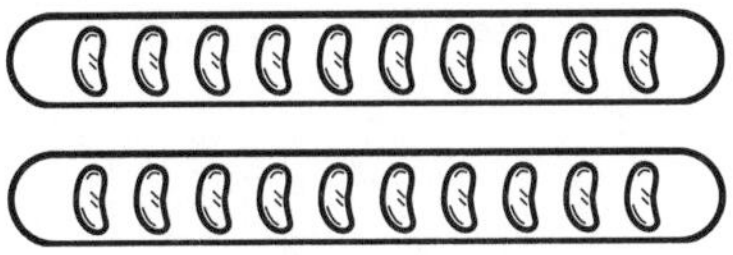 ____ tens ____ ones = ____

6. 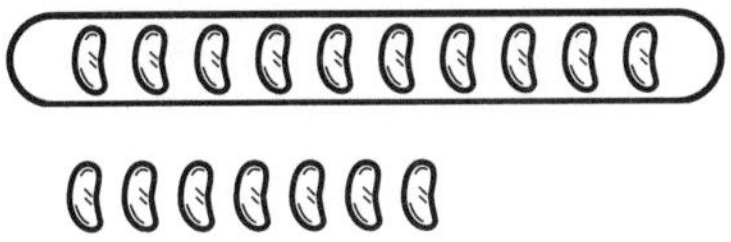 ____ ten ____ ones = ____

▶ **Problem Solving**

7. Joe has 10 toy cars. Mark gives him 2 more. How many toy cars does Joe have in all?

____ cars

ON MY OWN P71

Name _____

Tens and Ones to 50

LESSON 13.3

Write how many.

1.  32

2. _____

3. 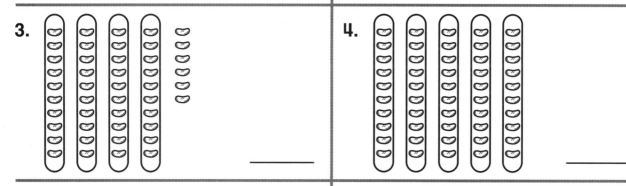 _____

4. _____

5.  _____

6. _____

▶ **Problem Solving**

Circle the box that has more.

7.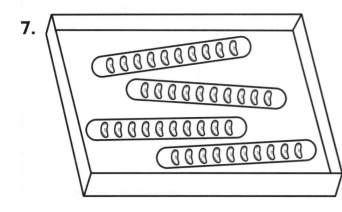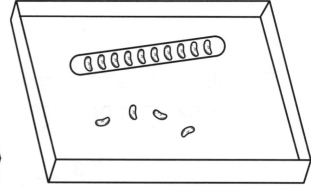

Name _____

LESSON 13.4

Tens and Ones to 80

Write the number.

1. 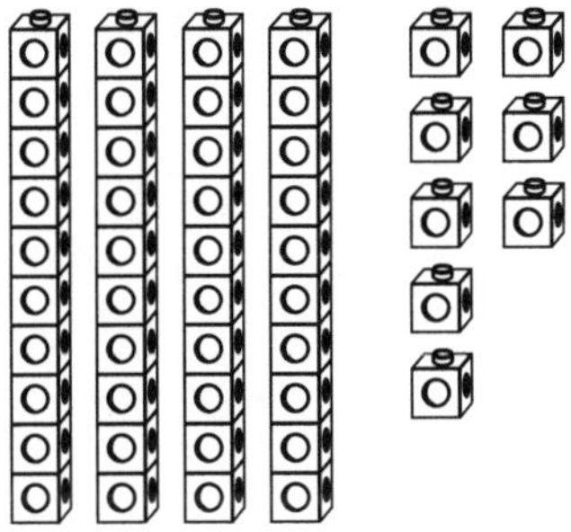 48

2. _____

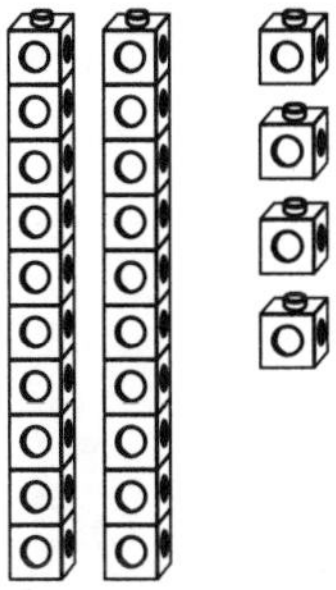

3. _____

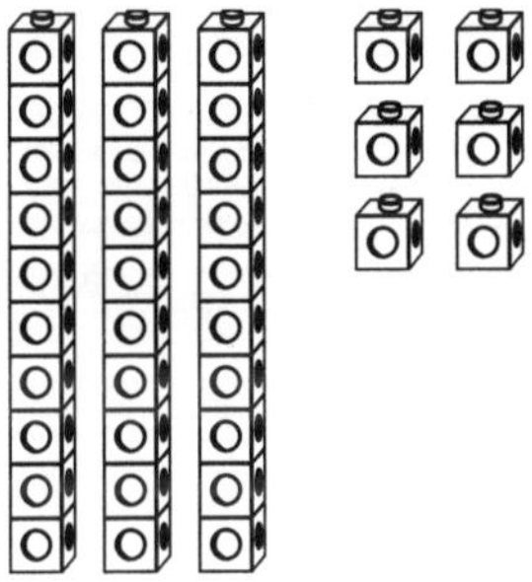

4. _____

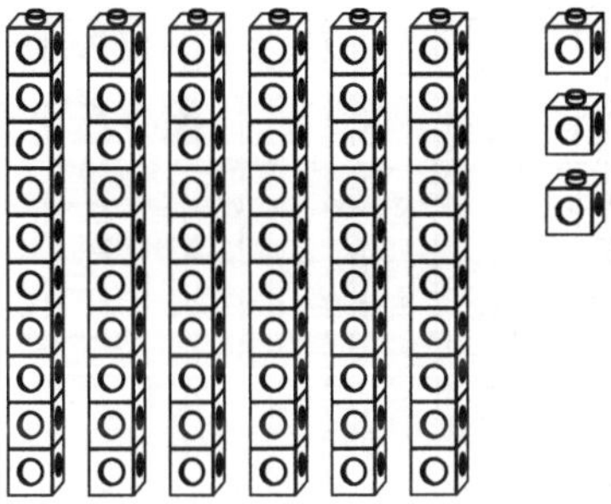

5. _____

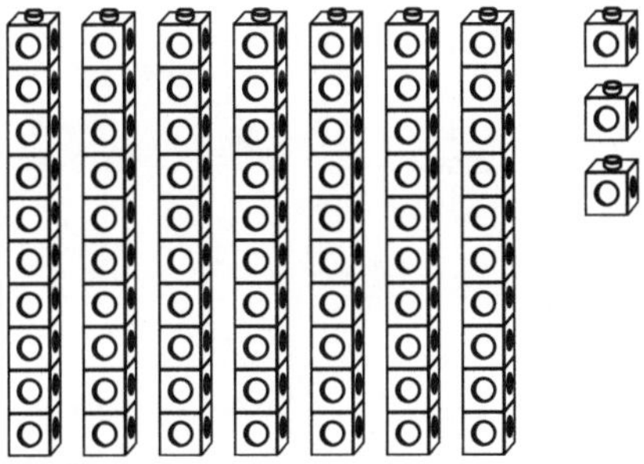

6. _____

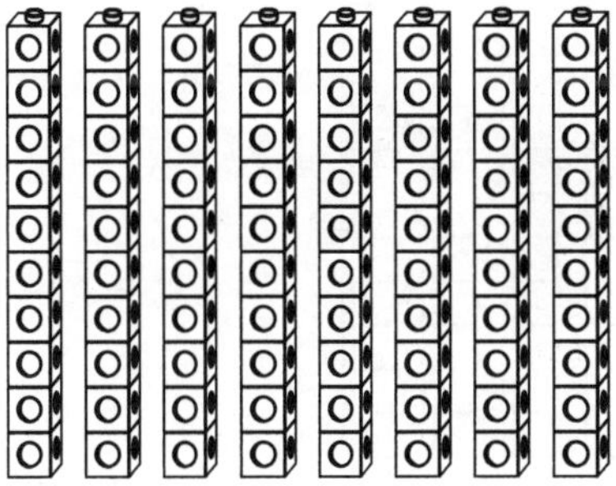

▶ **Problem Solving**

Write the number.

7. Mary has 1 ten and 3 ones. Joe has 3 tens and 1 one. Write the number that tells how many each one has.

Mary _____   Joe _____

ON MY OWN   P73

Name _____

LESSON 13.5

Tens and Ones to 100

Write the number.

1. 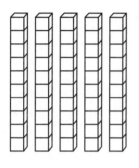 51

2. 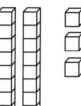 ____

3. 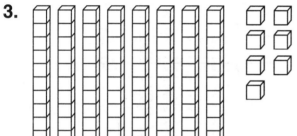 ____

4. 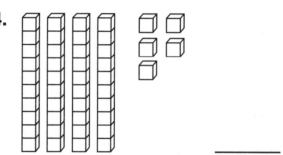 ____

5. 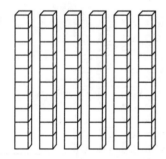 ____

6. 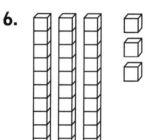 ____

▶ **Problem Solving**

Draw a picture to solve.

7. Kenda picked apples. She had 3 groups of 10 apples and 3 left over. How many apples did she have in all?

____ apples

Name _____

LESSON 13.6

Estimating 10

Circle the better estimate.

1.

   (more than 10)
   fewer than 10

2.

   more than 10
   fewer than 10

3.

   more than 10
   fewer than 10

4.

   more than 10
   fewer than 10

▶ **Problem Solving**

5. Matt has more than 15 but fewer than 20 oranges. Write the numbers that tell how many oranges Matt could have.

_____ _____ _____ _____

ON MY OWN P75

Ordinals

Match.

1. first   third   fifth   seventh   ninth   eleventh

second   fourth   sixth   eighth   tenth   twelfth

2. first   third   fifth   seventh   ninth   eleventh

second   fourth   sixth   eighth   tenth   twelfth

▶ **Problem Solving**

Circle the answer.

3. Which animal is first?

   rabbit   dog   duck   turtle

4. Which animal is third?

   rabbit   dog   duck   turtle

Name _____

LESSON 14.2

Greater Than

▶ Vocabulary

Write the numbers. Circle the **greater** number.

1.   (50)
     25

2.   _____
     _____

3.   _____
     _____

4.   _____
     _____

▶ Problem Solving

5. Bill has 2 dogs. Marc has more dogs than Bill. Circle the boy that is Marc.

6. The bus has 21 people on it. The plane has 40 people on it. Circle the one with the greater number on it.

ON MY OWN P77

Name _____

LESSON 14.3

Less Than

▶ **Vocabulary**

Write the numbers. Circle the number that is **less**.

1. 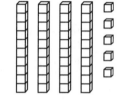   45
   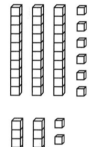   (43)

2. _____
   _____

3. _____
   _____

4. _____
   _____

▶ **Problem Solving**

5. Circle the one who has the most pennies.

   Jess   Sue   Pat   Bill

6. Mark an **X** on the one who has fewer pennies than Jess.

P78 **ON MY OWN**

Name _____

LESSON 14.4

Before, After, Between

▶ **Vocabulary**

33 is just **before** 34.

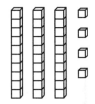

34 is **between** 33 and 35.

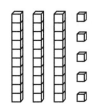

36 is just **after** 35.

Complete the tables.

before		after
1. 44	45	46
2. ___	60	___
3. ___	87	___
4. ___	23	___
5. ___	59	___
6. ___	11	___
7. ___	36	___
8. ___	98	___
9. ___	70	___

	between	
10. 46	___	48
11. 83	___	85
12. 19	___	21
13. 60	___	62
14. 55	___	57
15. 97	___	99
16. 78	___	80
17. 25	___	27
18. 89	___	91

▶ **Problem Solving**

Write the numbers in order.

19. | 25 | 26 | 24 |

___ ___ ___

20. | 55 | 53 | 54 |

___ ___ ___

ON MY OWN   P79

Name _____

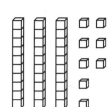

Order to 100

▶ **Vocabulary**

Write the numbers.
Then write them in order from **least** to **greatest**.

1. 8   28   15   38
    8   15   28   38

Write the numbers in order from least to greatest.

2. 57   59   53   55
    53   ____ ____ ____

3. 32   52   62   42
   ____ ____ ____ ____

4. 22   40   39   54
   ____ ____ ____ ____

5. 61   58   41   73
   ____ ____ ____ ____

▶ **Problem Solving**

6. Write three numbers that are
   between 52 and 57
   and that come after 53.   ____ ____ ____

LESSON 15.1

Counting by Tens

Count by tens. Write how many.

1.

   10   20   30   40   50   60

2.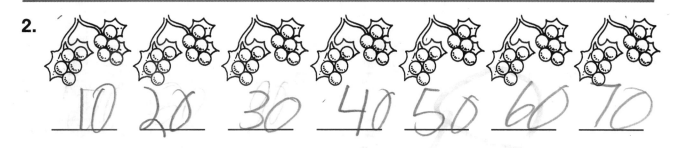

   10   20   30   40   50   60   70

▶ Problem Solving

Count by tens to fill in the table.
Use the table to answer the questions.

Megan has 20 pennies on Sunday.
She saves 10 pennies every day for one week.

Sunday	Monday	Tuesday	Wednesday	Thursday	Friday	Saturday
20	30	40	50	60	70	80

3. How many pennies does she have on Tuesday?

   _60_ pennies

4. How many pennies does she have on Friday?

   _70_ pennies

5. How many pennies does she have on Saturday?

   _80_ pennies

Name _____

LESSON 15.2

Counting by Fives

1. Write the missing numbers.
   Count by fives. Color those boxes red.
   Count by tens. Color those boxes blue.

1	2	3	4	5	6	7	8	9	10
11	12	13	14	15	16	17	18	19	20
21	22	23	24	25	26	27	28	29	30
31	32	33	34	35	36	37	38	39	40
41	42	43	44	45	46	47	48	49	50
51	52	53	54	55	56	57	58	59	60
61	62	63	64	65	66	67	68	69	70
71	72	73	74	75	76	77	78	79	80
81	82	83	84	85	86	87	88	89	90
91	92	93	94	95	96	97	98	99	100

▶ **Problem Solving**

2. Circle the numbers that are greater than 50.

   (80)   30   (90)   40   20

Name _____

LESSON 15.3

Counting by Twos

Count by twos. Write the missing numbers.

1.

1	2	3	4	5	6	7	8	9	10
11	12	13	14	15	16	17	18	19	20
21	22	23	24	25	26	27	28	29	30
31	32	33	34	35	36	37	38	39	40
41	42	43	44	45	46	47	48	49	50
51	52	53	54	55	56	57	58	59	60
61	62	63	64	65	66	67	68	69	70
71	72	73	74	75	76	77	78	79	80
81	82	83	84	85	86	87	88	89	90
91	92	93	94	95	96	97	98	99	100

2. My number is between 40 and 50. It is 2 more than 45. What is my number?

47

3. My number is between 80 and 90. It is 2 more than 87. What is my number?

89

Harcourt Brace School Publishers

ON MY OWN P83

Name_____

LESSON 15.4

Even and Odd Numbers

▶ **Vocabulary**

Circle **even** or **odd**.

1.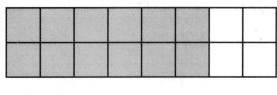

   12   even   odd

2.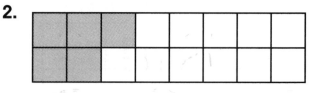

   5   even   odd

Color the squares to show each number. Circle **even** or **odd**.

3.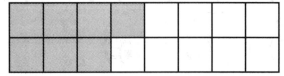

   7   even   (odd)

4.

   4   even   odd

5.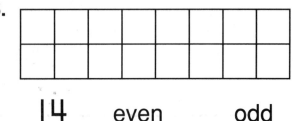

   14   even   odd

6.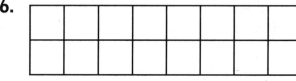

   9   even   odd

▶ **Problem Solving**

Circle **even** or **odd**.

7. Ann has 2 🚲.
   Each has 2 ⊙.
   Does she have an even or odd number of ⊙?

   even   odd

P84 **ON MY OWN**

Name _____

LESSON 16.1

Pennies and Nickels

▶ **Vocabulary**

1. Circle the **penny**.
2. Cross out the **nickel**.

Count. Write the amount.

3.

  __1__ ¢, __2__ ¢, __3__ ¢ | 3 | ¢

4.

  __1__ ¢, __2__ ¢, __3__ ¢, __4__ ¢, __5__ ¢, __6__ ¢ | 6 | ¢

5.

  __5__ ¢, __10__ ¢ | 10 | ¢

6.

  __5__ ¢, __10__ ¢, __15__ ¢ | 15 | ¢

▶ **Problem Solving**

Circle the greater amount.

7.

ON MY OWN   P85

Name _____

LESSON 16.2

Pennies and Dimes

▶ **Vocabulary**

1. Circle the **penny**.
2. Cross out the **dime**.

Count by tens. Write the amount.

3.
   __10__¢, __20__¢   __20__¢

4.
   __10__¢, __20__¢, __30__¢, __40__¢   __40__¢

5.
   __10__¢, __20__¢, __30__¢, __40__¢, __50__¢   __50__¢

6.
   __10__¢, __20__¢, __30__¢   __30__¢

7.
   __10__¢, __20__¢, __30__¢, __40__¢, __50__¢, __60__¢, __70__¢   __70__¢

▶ **Problem Solving**

Circle the least amount.

8.

P86 **ON MY OWN**

Name _____

LESSON 16.3

Counting Collections of Nickels and Pennies

Count by fives. Then count on by ones.
Write the amount.

1.   8 ¢

2.   11 ¢

3.   22 ¢

4.   26 ¢

▶ **Problem Solving**

Mark an **X** on the greater amount.

5.

6.

ON MY OWN   P87

Name_____

LESSON 16.4

Counting Collections of Dimes and Pennies

Count by tens. Then count on by ones.
Write the amount.

1.   31 ¢

2.   22 ¢

3.     53 ¢

4.   41 ¢

5.   14 ¢

▶ **Problem Solving**

6. Lucy wants to buy a paint set.
   It costs 43¢. Circle the coins Lucy needs.

ON MY OWN

Name _____

LESSON 16.5

Problem Solving • Choose the Model

Which two groups in each row add up to the amount on the tag? Color them.

1.

2.

3.

4.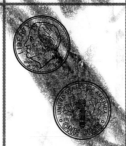

▶ **Problem Solving**

5. You have

     .

Circle the toy you can buy.

Name_____

LESSON 17.1

Trading Pennies, Nickels, and Dimes

Trade for nickels and dimes. Use the fewest coins.
Draw how many you have.

1.

| 10¢ | 10¢ | 10¢ | 5¢ |

2.

| 10 | 10 | 5 |

3.

| 5 | 5 | 5 |

4.

| 5 | 5 |

▶ **Problem Solving**

5. Wayne bought a watch.
He used 4 nickels.

Did he use the fewest
coins that equal 20¢? No

Show the price using
the fewest coins.

| 10 | 10 |

P90 **ON MY OWN**

Equal Amounts

Show the amount in 2 ways.
Circle the way that uses the fewest coins.

1.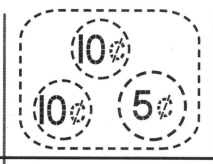

2.

3.

▶ **Problem Solving**

4. Lance needs 30¢ to buy a notebook.
   Mark an **X** on the groups that do not equal 30¢.

Which way uses the fewest coins to show 30¢? Circle it.

Name _____

LESSON 17.3

How Much Is Needed?

Circle the coins you need. Use the fewest coins.

1.

2.

3.

4.

▶ **Problem Solving**

5. Ruben has 15 ¢. He wants to trade his pennies for other coins. Show the fewest coins he can have.

Name _____

Ordering Months and Days

LESSON 18.1

			FEBRUARY			
Sunday	Monday	Tuesday	Wednesday	Thursday	Friday	Saturday
1	2	3	4	5	6	7
8	9	10	11	12	13	14
15	16	17	18	19	20	21
22	23	24	25	26	27	28

Write the days of the week in order.

1. _____
2. _____
3. _____
4. _____
5. _____
6. _____
7. _____

Color the Tuesdays .

Problem Solving

9. The ball game is on the day before Saturday. Write the day.

_____

10. _____day is in the last ____ the year. Circle the ____

December

Name_____

LESSON 18.2

Reading the Calendar

Fill in the calendar for next month.
Use the calendar to answer the questions.

_____

Sunday	Monday	Tuesday	Wednesday	Thursday	Friday	Saturday

1. On what day does the month end? _____

2. What is the date of the first Sunday? _____

3. What is the date of the first Friday? _____

▶ **Problem Solving**

4. Mike's birthday is on May 7. This year May 6 is on Thursday. On what day of the week is Mike's birthday? _____

P96 **ON MY OWN**

Name _____

LESSON 18.3

Ordering Events

Draw something special you do on each of these days.

Saturday night

Sunday morning

Monday afternoon

▶ **Problem Solving**

1. Circle the month that comes before December.

   January   November   February

2. Write the month that comes after October. _____

ON MY OWN   P97

Name _____

LESSON 18.4

Estimating Time

Circle the one that takes longer to do.

1.

2.

3.

▶ **Problem Solving**

4. Tyler and Sarah live next to each other. Tyler rides his bike home from school. Sarah walks home. Does it take more time for Tyler or Sarah to get home?

_____

P98 ON MY OWN

Name _____

Reading the Clock

LESSON 19.1

Use your clock. Show the time.
Write the time two ways.

1.
   __1__ o'clock
   1:00

2.
   __5__ o'clock
   5:00

3.
   __9__ o'clock
   9:00

4.
   __3__ o'clock
   3:00

5.
   __12__ o'clock
   12:00

6.
   __4__ o'clock
   4:00

▶ **Problem Solving**

7. It is 7 o'clock. Jenny has to go to bed in one hour.

 Yes

What time does Jenny go to bed? Write the time two ways.

__6__ o'clock
8:00

ON MY OWN P99

Name_____

LESSON
19.2

Hour

Write the time.

1.
12:00

2.
5:00

3.
1:00

4.
9:00

5.
3:00

6.
7:00

7.
2:00

8.
6:00

9.
4:00

▶ **Problem Solving**

10. Write the time on the clock so that it shows 1 hour later than 6 o'clock.

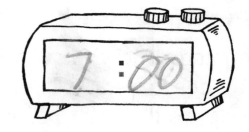
7:00

Name _____

LESSON 19.3

Hour

▶ **Vocabulary**

Draw the **hour hand** and the **minute hand**.

1.

2.

3.

4.

5.

6.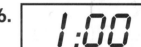

7.

8.

9.

▶ **Problem Solving**

10. Jonathan eats dinner at 6 o'clock. He goes to bed 2 hours later. Write the time he goes to bed.

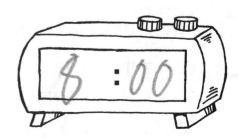

ON MY OWN

Name _____

LESSON 19.4

Half Hour

Write the time.

1.
5:30

2.
7:30

3.
1:30

4.
2:30

5.
9:30

6.
3:30

7.
12:30

8.
10:30

9.
6:30

▶ **Problem Solving**

10. The movie starts at 7:30.
    It lasts for two hours.
    What time is the movie over?
    Circle the clock that shows
    when the movie is over.

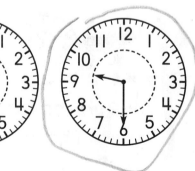

Name _____

Problem Solving • Act It Out

LESSON 19.5

About how long would it take? Circle your estimate.
Then act it out to see if you are right.

1. put 10 chairs in a circle

(more than a minute)
less than a minute

2. put a stamp on a letter

more than a minute
less than a minute

3. open a door

more than a minute
less than a minute

4. write 10 spelling words

more than a minute
less than a minute

5. read a big book

more than a minute
less than a minute

6. sharpen a pencil

more than a minute
less than a minute

ON MY OWN P103

Name _____

Using Nonstandard Units

LESSON 20.1

Estimate. Then use to measure.

1.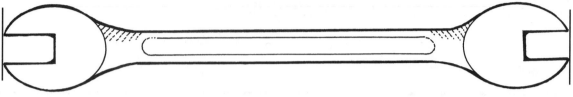

   Estimate   about _____   Measure   about _____

2.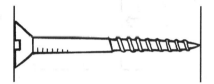

   Estimate   about _____   Measure   about _____

3.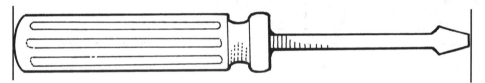

   Estimate   about _____   Measure   about _____

4.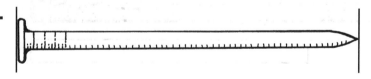

   Estimate   about _____   Measure   about _____

▶ **Problem Solving**

5. Use and to measure.
   Circle which way uses more.

P104   **ON MY OWN**

Name _____

LESSON 20.2

Measuring in Inch Units

▶ Vocabulary

1. Circle the pencil that is 1 **inch** long.

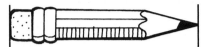   

Count the inch units. Write how many inches long.

2.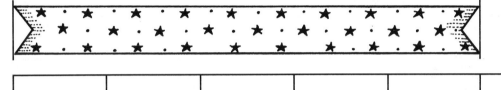

_5_ inches

3.

____ inches

4.

____ inches

▶ Problem Solving

5. One 📎 is about 1 inch long. About how many inches long are 3 📎? Draw a line to show.

about ____ inches long

ON MY OWN P105

Name _____

Using an Inch Ruler

You will need: objects, an inch ruler

Estimate. Then use an inch ruler to measure.

Object	Estimate	Measure
1. ![book]	about ____ inches	about ____ inches
2. ![tissues]	about ____ inches	about ____ inches
3. ![pencil]	about ____ inches	about ____ inches
4. ![paint set]	about ____ inches	about ____ inches

▶ **Problem Solving**

5. Measure each chain. If you joined the two chains, how long would the new chain be?

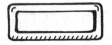 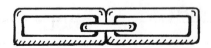   ____ inches

Name _____

LESSON 20.4

Measuring in Centimeter Units

▶ **Vocabulary**

Measure. Circle the rope that is 1 **centimeter** long.

How many centimeters long?
Count the centimeter units. Write how many.

1.

   _3_ centimeters

2.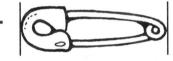

   ____ centimeters

3.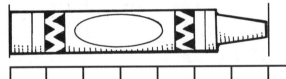

   ____ centimeters

4.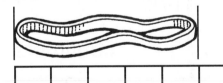

   ____ centimeters

▶ **Problem Solving**

5. Draw a 10 centimeters long.

ON MY OWN P107

Name_____

LESSON 20.5

Using a Centimeter Ruler

You will need: objects, a centimeter ruler

Estimate. Then use a centimeter ruler to measure.

Object	Estimate	Measure
1. (eraser)	about ____ centimeters	about ____ centimeters
2. (lunchbox)	about ____ centimeters	about ____ centimeters
3. (marker)	about ____ centimeters	about ____ centimeters
4. (notepad)	about ____ centimeters	about ____ centimeters

▶ **Problem Solving**

5. Use a centimeter ruler. Measure the sides of the rectangle.

Side A

Side B

Side A _____ centimeters

Side B _____ centimeters

P108 **ON MY OWN**

Name _____

LESSON 21.1

Using a Balance

▶ **Vocabulary**

1. Circle the object that is **heavier**.

2. Circle the object that is **lighter**.

You will need: a ⚖. Work with a partner. Find each object. Is the object heavier or lighter than a bottle of glue?

Estimate. Write **H** for heavier. Write **L** for lighter. Then use a ⚖ to measure. Write H or L.

Object	Estimate	Measure
3. ✏️		
4. 🗒️		
5. 📦		

▶ **Problem Solving**

6. Ashley has 2 cups. One is full of marbles. One is full of chalk. Circle the cup that is heavier.

ON MY OWN   P109

Name _____

LESSON 21.2

Using Nonstandard Units

You will need: a ⚖ and 🧊

Find each object. About how many 🧊 does it take to balance the scale? Estimate. Then measure.

Object	Estimate	Measure
1. (pen)	about ____ 🧊	about ____ 🧊
2. (notepad)	about ____ 🧊	about ____ 🧊
3. (Crayons)	about ____ 🧊	about ____ 🧊
4. (watch)	about ____ 🧊	about ____ 🧊

▶ **Problem Solving**

Look at your measures for the objects.

5. Draw a picture of the heaviest object.

6. Draw a picture of the lightest object.

P110 **ON MY OWN**

Name _____

Measuring with Cups

About how many cups of rice does each object hold? Estimate. Then measure.

Object	Estimate	Measure
1. (jar)	about ____ cups	about ____ cups
2. (bowl)	about ____ cups	about ____ cups
3. (measuring cup)	about ____ cups	about ____ cups
4. (Quart jar)	about ____ cups	about ____ cups

▶ **Problem Solving**

Look at the container.
Draw a container beside it that holds more.

5.

6.

ON MY OWN

Name _____

LESSON 21.4

Temperature
Hot and Cold

Circle the picture that shows something hot.

1.

2.

Circle the picture that shows something cold.

3.

4.

▶ **Problem Solving**

5. Draw something hot.

6. Draw something cold.

Name _____

Equal and Unequal Parts of Wholes

▶ **Vocabulary**

1. Circle the figure that shows **equal parts.**

2. Mark an **X** on the one that does not show equal parts.

Circle the figures that show equal parts.

3.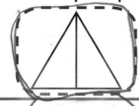

4.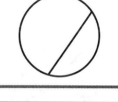

5.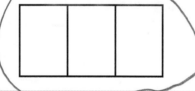

6.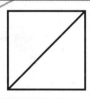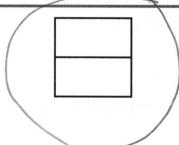

▶ **Problem Solving**

Draw lines to show where you would cut this cake.

7. Each child wants an equal share.

ON MY OWN P113

Name **LePriece**

LESSON 22.2

Halves

Find the figures that show halves. Color $\frac{1}{2}$ red.

1.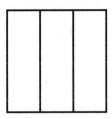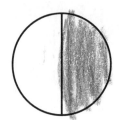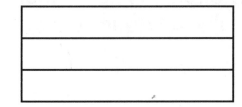

2.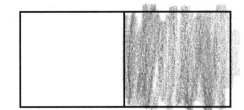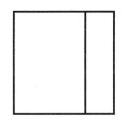

3.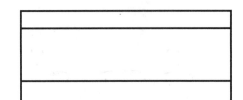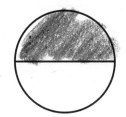

4.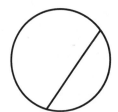

▶ **Problem Solving**

5. Two rabbits shared 4 carrots. Each ate one half of the 4 carrots. How many carrots did each rabbit eat? __2__ carrots

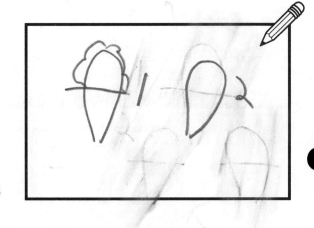

P114 **ON MY OWN**

Name _____

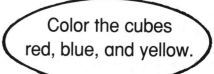

Most Likely

You will need: 1 bag, 6 red cubes, 5 blue cubes, and 1 yellow cube

Put the cubes into the bag. Take out one cube.

Make a tally mark on the table to show which color you got.

Put the cube back into the bag. Shake. Make a prediction. If you do this 9 more times, which color do you think you will get most often? Circle that color.

Color the cubes red, blue, and yellow.

red **blue** **yellow**

Do this 9 more times.

Make a tally mark each time. Count the tally marks for each color.

Write the totals.

	Tally Marks	Total
red		
blue		
yellow		

▶ **Problem Solving**

Can you take a △ out of the ?

Circle **Yes** or **No.**

    Yes   No

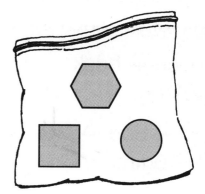

Name_____

LESSON 23.4

Tallying Events

You will need: 1 bag, 5 red cubes, 3 blue cubes

Color the cubes. Predict which color cube you will get more often. Circle that cube.

Try it. Put all the cubes into the bag. Color the cubes in the table. Take out a cube. Make a tally mark.

Put the cube back into the bag.

Shake the bag. Do this 10 times.

red   blue

	Tally Marks	Total
red		
blue		

Try again with these cubes.

You will need: 1 bag, 2 red cubes, 7 blue cubes

Color the cubes. Predict which color cube you will get more often. Circle that cube.

Try it. Color the cubes in the table. Put all the cubes into the bag. Take out a cube. Make a tally mark.

Put the cube back into the bag.

Shake the bag. Do this 10 times.

red   blue

	Tally Marks	Total
red		
blue		

Name _____

LESSON 24.1

Picture Graphs

bear	🐻	🐻	🐻	🐻	🐻
car	🚗	🚗			
doll	🪆	🪆	🪆	🪆	

Count the toys. Draw ◯ to fill in the graph.

	Our Favorite Toys				
bear	◯	◯	◯	◯	◯
car					
doll					

Use the graph to answer the questions.

1. How many cars are there? _____

2. How many more bears than cars are there? _____
   _____

3. Are there more dolls or cars? -----------------------
   _____

 Problem Solving

Look at the graph.

4. Write a number sentence that tells how many bears and dolls there are.

   ____ ◯ ____ = ____

   ____ bears and dolls

Horizontal Bar Graphs

A group of children voted for their favorite sports. Write how many tally marks.

Favorite Sport		Total
soccer	⊮ ll	7
softball	⊮	
football	lll	

Color the graph to match the tally marks.

Favorite Sport								
soccer	▓	▓	▓	▓	▓	▓	▓	
softball								
football								
	0	1	2	3	4	5	6	7

Use the graph to answer the questions.

1. How many children voted for soccer? ____

2. How many more voted for soccer than for football? ____

3. How many more voted for soccer than for softball? ____

▶ **Problem Solving**

Look at the graph.

4. Write a number sentence that tells how many more children like soccer than like football.

____ ◯ ____ = ____

____ more children

Name _____

LESSON 25.1

Doubles Plus One

1. Circle the **doubles** fact.

   4 + 4 = 8   4 + 3 = 7   4 + 5 = 9

2. Circle the **doubles plus one** fact.

   4 + 4 = 8   4 + 3 = 7   4 + 5 = 9

Write the sums.

3. 3 + 3 = 6,   so 3 + 4 = __7__.

4. 8 + 8 = 16, so 8 + 9 = ____.

5. 6 + 6 = 12, so 6 + 7 = ____.

6. 2 + 2 = 4,   so 2 + 3 = ____.

7. 4 + 4 = 8,   so 4 + 5 = ____.

Write the sums.

8.   7   8   5   5   4   1
   +7   +9   +5   +6   +4   +1

▶ **Problem Solving**

Use counters to solve. Draw them.

9. Bill has 5 marbles. He finds 6 more marbles. How many marbles does he have in all?

   ____ marbles

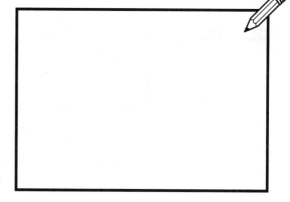

ON MY OWN P127

Name _____

LESSON 25.2

Doubles Minus One

1. Circle the **doubles** fact.

   3 + 4 = 7   3 + 3 = 6   3 + 2 = 5

2. Circle the **doubles minus one** fact.

   3 + 4 = 7   3 + 3 = 6   3 + 2 = 5

Write the sums.

3. 4 + 4 = 8 | 6 + 6 = ___ | 3 + 3 = ___
   4 + 3 = 7 | 6 + 5 = ___ | 3 + 2 = ___

4. 5 + 5 = ___ | 9 + 9 = ___ | 7 + 7 = ___
   5 + 4 = ___ | 9 + 8 = ___ | 7 + 6 = ___

5. 9   2   7   8   6   5
   +8   +3   +8   +9   +7   +6

6. 7   5   4   4   8   3
   +6   +4   +3   +5   +7   +2

▶ **Problem Solving**

7. Sue has 4 pennies. She finds 5 more. How many does she have in all?

   ___ pennies

P128 **ON MY OWN**

Name ______________________________

LESSON 24.3

Vertical Bar Graphs

Write how many tally marks. Color the graph to match the tally marks.

Where We Went for Vacation		Total
beach	𝍸 𝍷𝍷𝍷𝍷	9
city	𝍸	
farm	𝍷𝍷𝍷𝍷	

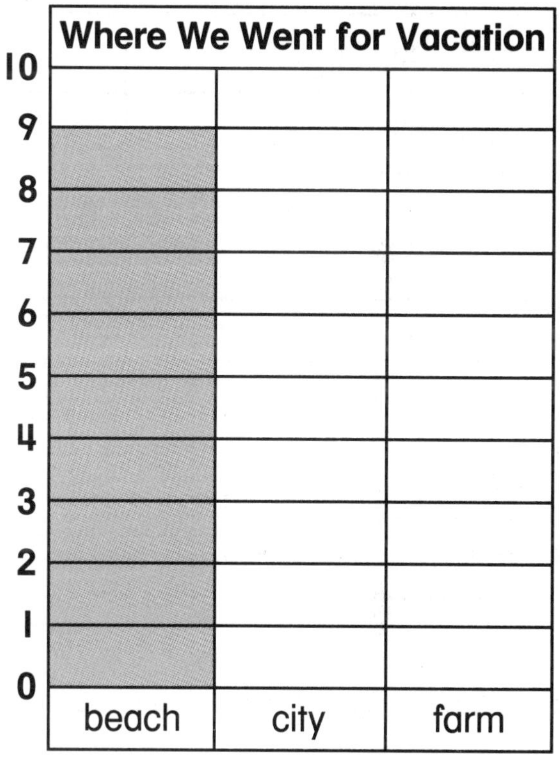

Use the graph to answer the questions.

1. How many children went to the beach? 9

2. How many children went to the farm? ______

3. How many more children went to the beach than to the farm? ______

Problem Solving

4. Circle the question you can answer by reading the graph.

How many children went to the mountains?

How many children went to the city?

Name ______________________

LESSON 24.4

Problem Solving • Make a Graph

Ask 10 classmates to choose their favorite color.

1. Make a tally mark for each choice. Then write how many.

2. Fill in the graph. First write the title. Then color the graph to match the tally marks.

Our Favorite Color		Total
red		
blue		
green		
yellow		

	______________			
8				
7				
6				
5				
4				
3				
2				
1				
0	red	blue	green	yellow

Use the graph to answer the questions.

3. Which color do the most children like best? ______

4. How many children like yellow the best? ______

5. Write a question someone can answer by reading this graph.

____

____

Name _____

LESSON 25.5

Problem Solving • Make a Model

Use counters to solve. Draw them.

1. Robert has 7 pencils. Sara has 1 fewer than Robert. How many pencils do they have in all?

   __13__ pencils

2. Lisa has 4 dolls. Kara has two times as many. How many dolls do they have in all?

   _____ dolls

3. John had 14 boats. He gave some of them away. He has 7 left. How many did he give away?

   _____ boats

▶ **Problem Solving**

4. I have 8 toy cars. My friend has the same number. How many cars do we have in all?

   _____ cars

ON MY OWN   P 131

Name _____

LESSON 26.1

Make a 10

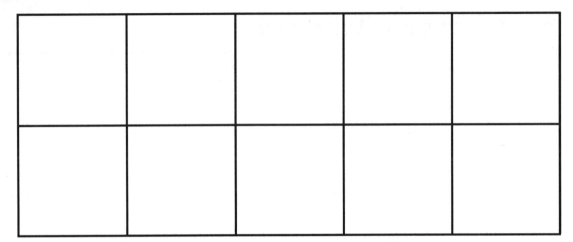

Use counters and the 10-frame. Start with the greater number. Make a 10. Then add.

1.
   3   8   5   3   9   6
   +9   +5   +9   +8   +7   +8

   12

2.
   8   9   7   4   3   6
   +7   +3   +9   +8   +9   +7

3.
   8   7   3   3   9   7
   +6   +4   +8   +9   +6   +5

▶ **Problem Solving**

4. Mack had 9 pencils. His dad gave him 2 more. How many pencils does he have in all? _____ pencils

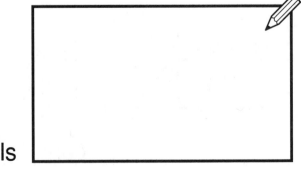

P132 ON MY OWN

Name _____

LESSON 26.2

Adding Three Numbers

Circle names for 10 or doubles. Then add.

1.
```
  1        3        6        7        2
 (8)       3        5        7        8
+(8)      +6       +4       +2       +5
 ---      
 17
```

2.
```
  3        1        8        7        5
  7        7        2        4        5
 +2       +7       +1       +3       +1
```

3.
```
  9        2        9        6        4
  2        8        3        6        3
 +1       +6       +1       +3       +4
```

▶ **Problem Solving**

4. Jan has 8 yellow leaves, 2 red leaves, and 6 brown leaves. How many leaves does she have in all?

_____ leaves

Name _____

Sums and Differences to 14

LESSON 26.3

Use Workmat 2 and counters. Add or subtract.

1.
   9   10   8   11   5   10
+1   −1   +3   −3   +5   −5
10   9

2.
   8   13   7   14   6   12
+5   −5   +7   −7   +6   −6

3.
   9   14   4   10   9   12
+5   −5   +6   −6   +3   −3

4.
   9   11   8   14   5   9
+2   −2   +6   −6   +4   −4

▶ **Problem Solving**

5. Joey has 14 stickers. He gives 6 to Randy. How many stickers does Joey have left?

_____ stickers

P134 **ON MY OWN**

Name _____

LESSON 26.4

Sums and Differences to 18

Write the sum and difference for each pair.

1.
    9   18   8   14   6   12
    +9 −9   +6 −6   +6 −6
    18   9

2.
    8   11   9   13   7   13
    +3 −3   +4 −4   +6 −6

3.
    9   16   6   11   6   14
    +7 −7   +5 −5   +8 −8

4.
    6   15   9   12   6   13
    +9 −9   +3 −3   +7 −7

▶ **Problem Solving**

5. I found 17 shells. I gave away 9. How many shells do I have left?

_____ shells

ON MY OWN P135

Name_____

Counting Equal Groups

Use counters. Draw them. Write how many in all.

1. Make 2 groups. Put 3 counters in each group.

   How many in all? __6__

2. Make 3 groups. Put 2 counters in each group.

   How many in all? _____

3. Make 2 groups. Put 4 counters in each group.

   How many in all? _____

4. Make 4 groups. Put 3 counters in each group.

   How many in all? _____

▶ **Problem Solving**

5. Jessica has 3 dogs. She gave them 2 bones each. How many bones did Jessica need?

   _____ bones

Name _____

LESSON 27.2

How Many in Each Group?

Use counters. Draw them. Write how many in each group.

1. Use 10 counters. Make 5 equal groups.

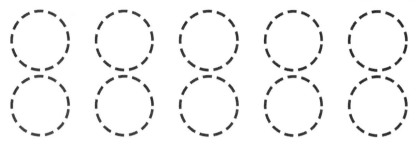

How many in each group? __2__

2. Use 12 counters. Make 2 equal groups.

How many in each group? _____

▶ **Problem Solving**

3. Dylan gave 4 pictures to his brother and sister. He gave each the same number. How many pictures did he give to each?

   sister _____ brother _____

4. Jenny's grandmother gave 12 cookies to 3 children. She gave each child the same number of cookies. How many cookies did each child get?

   _____ cookies

ON MY OWN P137

Name_____

LESSON 27.3

How Many Groups?

Use counters. Draw them. Write how many groups.

1. Use 12 counters. Put 3 in each group.

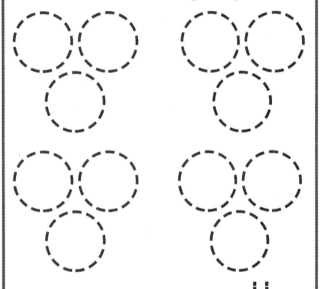

How many groups? __4__

2. Use 8 counters. Put 2 in each group.

How many groups? _____

▶ **Problem Solving**

3. The teacher had 6 apples. She put 2 apples on each plate. How many plates did she use?

_____ plates

4. Jason had 10 sugar cubes. He gave 5 to each horse. How many horses did he feed?

_____ horses

P138 **ON MY OWN**

Name _____

Problem Solving • Draw a Picture

Draw a picture to solve each problem.

1. There are 2 children.
   Each child has 3 balloons.
   How many balloons are there?

   __6__ balloons

   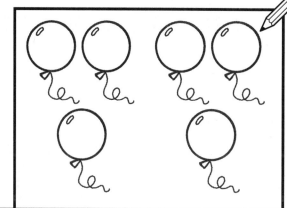

2. We have 3 pennies.
   We get 2 more.
   How many pennies
   in all?

   _____ pennies

3. There are 4 baskets.
   Each basket has 3 eggs in it.
   How many eggs are there?

   _____ eggs

4. There are 6 crayons.
   Each child gets 2 crayons.
   How many children
   get crayons?

   _____ children

LESSON 27.4

ON MY OWN

Name _____

LESSON 28.1

Adding and Subtracting Tens

 Vocabulary

Circle the Workmat that shows 4 **tens**.

tens	ones
	□ □
	□ □

tens	ones
▌▌▌▌	

Add or subtract. Use tens and Workmat 3.

1.
```
  40        70        50        30        30
+ 20      + 10      + 20      + 20      + 10
----      ----      ----      ----      ----
  60
```

2.
```
  40        90        70        80        60
- 30      - 30      - 30      - 20      - 40
----      ----      ----      ----      ----
```

▶ **Problem Solving**

3. Ann had 40¢ to buy a pencil. The pencil cost 30¢. How much money does Ann have left?

_____ ¢

P140 **ON MY OWN**

Name _____

LESSON 28.4

Reasonable Answer

Circle the answer that makes sense.

1. Chris had 17 books. He gave 12 to the library. How many books does Chris have now?

   55 books
   (5 books)
   555 books

2. There were 35 children at the zoo. 12 children went home. How many children are left?

   650 children
   76 children
   23 children

3. Steve had 13 stickers. Then he bought 20 more. How many stickers does he have now?

   5 stickers
   133 stickers
   33 stickers

4. Rose saw 6 squirrels on the fence. Then she saw 12 more in the grass. How many squirrels were there in all?

   18 squirrels
   6 squirrels
   118 squirrels

▶ **Problem Solving**

Circle the answer.

5. Juan has 35¢. He wants to buy a toy car that costs 22¢. How much money will he have left?

   35¢   22¢   13¢

6. Sandy had 45 pennies. Her mother gave her 23 more. How many pennies does Sandy have now?

   21¢   45¢   68¢

ON MY OWN   P143